Tricks of the Poultry Trade
Some Methods and Little Things Practiced Among the Initiated of the Craft

by Reese V. Hicks

with an introduction by Jackson Chambers

This work contains material that was originally published in 1912.

This publication is within the Public Domain.

*This edition is reprinted for educational purposes
and in accordance with all applicable Federal Laws.*

Introduction Copyright 2018 by Jackson Chambers

The World's Largest Selection of Vintage Poultry Books

www.VintagePoultry.com

Self Reliance Books

Get more historic titles on animal and stock breeding, gardening and old fashioned skills by visiting us at:

http://selfreliancebooks.blogspot.com/

Introduction

I am pleased to present yet another title on Poultry.

The work is in the Public Domain and is re-printed here in accordance with Federal Laws.

As with all reprinted books of this age that are intended to perfectly reproduce the original edition, considerable pains and effort had to be undertaken to correct fading and sometimes outright damage to existing proofs of this title. At times, this task is quite monumental, requiring an almost total "rebuilding" of some pages from digital proofs of multiple copies. Despite this, imperfections still sometimes exist in the final proof and may detract from the visual appearance of the text.

I hope you enjoy reading this book as much as I enjoyed making it available to readers again.

Jackson Chambers

INTRODUCTION

In recent years a wonderful revival has taken place in the poultry business. A tide has set in from city to country. All this has created an interest for poultry information and especially for "nigh-cuts" to wealth and health at the same time. Many so-called "systems," "processes" and "discoveries" have been heralded through the poultry journals and many of these "discoveries" sold at fancy prices. It is an old and true saying that there is nothing new under the sun. The majority of these much heralded "discoveries" have been known among poultrymen for years. Believing that a brief telling to them in a plain, practical way will be of help, is the excuse for this book. It is not claimed that these systems or discoveries are original with the writer of this pamphlet, but have been known to him and many others, as well as practiced by them, for years before many of these "discoverers" published them. There are good points in some of these systems, and, if carefully put in practice, will prove valuable to anyone.

The word "trick" suggests in itself something clever and, to the minds of many, means an action of "shady" character. This treatise will touch upon both sides, giving many little things—and some large ones—that go to make success in handling poultry. It will also tell of some things done that should be left undone—and their telling here does not mean, "Go thou and do likewise," but these things are only given to warn the unwary against them.

This book is submitted for your careful study, as the information given herein is valuable only when applied by a skillful hand and a knowing head.

REESE V. HICKS.

Topeka, Kansas, July 1, 1909.

THIRD REVISED EDITION.

This edition is brought out in response to a wide demand for the information given in this book. The entire subject matter has been carefully read and, where it needed, revised.

The Editor, Topeka, Kansas, March 5, 1912.

INFORMATION FOR THE BUYER

HOW TO GET A GOOD START.

There are two methods of getting a start in poultry.

First. Buy the stock and raise your own flock.

Second. Buy eggs or baby chicks and raise the foundation stock.

Each of these methods has its advantages, as well as disadvantages. Buying eggs is largely buying "sight unseen." Buying eggs is largely a gamble, because breeders do not know positively what will be the result of their matings.

The stock method is costlier than to start by the egg method, but, on the other hand, you can get a good sized flock the first year and thus get in the business a year earlier.

Avoid cheap stock and eggs unless the breeder gives you a very clear and good reason for selling cheap. Some good breeders offer special prices on eggs and stock after setting season.

It takes years to breed up a flock, and stock and eggs cannot be sold from such flocks except at good prices. Birds from wellbred flocks, with anything like good marking, are worth from $5.00 up, while eggs are worth from such flocks $2.00 per 15 up.

When you go to buy and want good stuff, sit down and write and tell the breeder you are going to buy from what you want. Ask him to send you stock mated to produce that, or else eggs from stock mated for that purpose. Many of the best fanciers get their best birds from matings that the average buyer would not consider the birds in the pens worthy of consideration.

HINTS ON VITALITY.

It is very important that you have highly vital birds when you select breeders. Have the males look their part—robust and thrifty—full of vim and energy. The females should look trim and neat in outline, and still showing vitality and strength. Comb and wattles are important indications of vitality, and should be well developed, consdering the breed, and of good color. Never use in breeding pens a bird that has sickly looking comb and wattles or these organs under size.

The eye is one of the most important indications, as it shows the life and vitality in the bird. Small eyes are to be avoided, as well as those havng a dull and sickly look to them.

The beak should have a reasonable curve and be of good size. Never breed from a bird with a small, straight, spindling beak.

Feet are good pointers and should be well developed and large.

In considering all these points, breed traits must be considered. What would be a large comb on a Plymouth Rock would be a small one on a Minorca, and large legs on a Leghorn would be small on a Plymouth Rock, etc.

Carriage or attitude is a good pointer on vitality. Never breed from a bird droopy and not alert.

Breadth of back is a sign of a strong, vigorous bird, and should be closely looked to in selecting males, unless a small, trim shape is desired in the breed. The head on a vigorous bird should not be "snakey," nor the neck extra long and slim for the breed.

The vigorous, strong birds cackle or crow louder, show a greater appetite and a more restless disposition than the weaker ones.

The vigorous bird is usually of brighter plumage and not so dull and lustreless.

TELLING THE AGE OF FOWLS.

Telling the age of old fowls is largely guesswork. The first year of a chicken's life can be very easily told. There are more long hairs on an old chicken and fewer pin feathers. On a young and growing chicken there are more undeveloped feathers. The breast and pelvic bones in a young chicken at the tip are tender and flexible and bend on slight pressure. After the bird has passed a year old, these bones become hard. The feet and legs of a young bird are smoother, cleaner and free from scaley leg, bumble foot and the toenails show no wear. While scaly leg, bumble foot, etc., do not always appear on a bird over a year old, and their appearances are a pretty sure sign of this age, yet their absence are not infallible signs. The breast bone, as well as the point of the two pelvic bones, frequently get their hardness in 8 to 14 months, owing to the breed. The smaller breeds get hard on breast and pelvic bones sooner than the large breeds, the Asiatics being the slowest in hardening these bones.

The amount of fat on the bird indicates the age, as old birds take fat more readily than young ones. The brightness of plumage is an indication of a young bird, as the faded, "scraggy" bird is nearly always an old one. Bad keeping and poor attention make birds appear much older than they really are—it ages them in usefulness as well as appearance. Frequently "off colored" feathers will appear on the neck and head of a hen or cock when they have passed their first or second year. There is also a sunken appearance around the eyes of birds past the second or third year. Old hens frequently have a "baggy" effect in the rear, caused by age, fat and heavy laying. A hen with this baggy, broken down condition in the rear is in effect old, as she has passed her day of usefulness, no matter whether 1 year old or 10.

The spurs are a good indication of age, especially on male birds. The spurs are said to be "set"—that is, become fixed to the leg bone at from 8 to 10 months, according to the breed, the small breeds "setting" their spurs earlier than the larger breeds. This test does not apply to hens so much as to male birds. A male spur is one-half to three-fourths on an inch long at 1 year old and round pointed. Older birds have not only longer, but larger, sharper and rougher spurs. Hens differ much in size of spurs, and the size of a hen's spur is not much of an indication as to age, except a large spur means age, but its absence does not mean youth.

A young pullet shows rose colored veins under the wings, which disappear n a matured bird. The old bird has coarser and drier skin.

In young chickens from two to four months old, the first feathers after the furze on the main wing and tail feathers are sharp pointed, while the adult feathers are round pointed. Some birds carry them only two or three months, depending on breed, conditions of growth, etc.

In ducks and geese, the windpipe is an almost infallible indication of age. Young ducks and geese have soft, flexible, cartilage in the windpipe, while an old bird's is considerably harder. The breast bone is also a sure indication with ducks and geese, as it is very pliable in young birds and becomes hardened in the old ones.

In telling the age of turkeys, the turkey does not molt its feathers as soon as a chicken, and a turkey under a year old will usually have sharp pointed feathers, while birds over that age will have rounded points.

In males of turkeys, the length of the beard indicates the youth or age of the bird. In Bronze turkeys, the shanks and toes of the matured fowls take on a pinkish color, while the young and yearlings have dark, nearly black, shanks.

TELLING THE AGE OF EGGS.

Expert handlers of eggs learn to tell cold storage eggs by the amount of evaporation they show. Eggs that have been in cold storage show much larger air cells, as the contents of the eggs shrink all the time. Cold storage eggs also show an unevenness in the contents on account of having laid a long time in one position. They show mold on the shells often. When boiling cold storage eggs with the shells on, they show a marked tendency to crack. The yellows in cold storage eggs break easily, and do not stand up whole like fresh eggs.

A very fair estimate of the age of an egg can be told by the means of its specific gravity. Eggs dry out constantly from the time laid, and the older they are the lower the specific gravity. It is true there is considerable difference in specific gravity of eggs when first laid, but the eggs low in specific gravity are usually pretty sure to be old, as the moisture has evaporated.

A simple method of telling the specific gravity is to use heavily salted water. For full description of this method, see illustration and directions under the heading of "Testing Eggs Before Setting," on later pages.

Eggs have a "bloom," or tint, that is familiar to all who handle eggs. This "bloom" is destroyed by washing or considerable handling, and also eggs kept on hand for some time naturally lose some of their bloom through contact with the air. Eggs that have been incubated under a hen have a slick, greasy appearance, and those incubated in an incubator do not show this, but the bloom is very much affected. The absence of the "bloom" on an egg is a pretty fair indication of age or usage.

Houses, Hoppers and Fixtures

LOCATION OF POULTRY PLANT.

An unfavorably located poultry plant is likely to be a failure. The first thing to avoid in locating a poultry house is a damp and swampy place, as dampness affects both young and old fowls. The ideal location is one sloping gently to the southeast. Avoid steep hillside for a poultry plant, as it means extra exertion in getting in and out with feed, cleaning, etc. Very shady locations are also bad, although an orchard is a fine place for stock to range in spring and summer, but the shade should not be over the buildings or over more than half of the yards and runs. Sunshine is necessary for proper sanitation.

THE HOUSE AND RUNS.

Simplicity is one of the most important things in your poultry house and fixtures, and is a stumbling block that nearly all beginners fall over when they build their poultry houses, etc. Have them simple and free from cumbering details. It takes time to clean complicated fountains, houses, hoppers, etc., and to insure reasonable results with a flock of poultry the houses, etc., must be cleaned often.

The modern idea in poultry house building is to get as close to nature as possible. In other words, give the poultry all the air and sunshine you can, providing, of course, against excessive sunshine and heat in the summer. The open front house, as it is called on account of the fact that the whole front, or a good portion of it, is open or at least closed only by a light curtain, is gaining more and more popularity. The true "open front" house has no curtain at all on one side. The open side should be towards the south. The "curtain front" house is an open front house, also, except that a curtain of muslin or some similar material is provided to drop down over the front in unfavorable weather and on some nights.

MODERN POULTRY HOUSES.

When you begin to build a house, you should make up a line of questions that are to be answered before you build, or rather answered in the building. A good list is as follows:

1. Location of house. 2. Size of house. 3. Height of house. 4. Style of roof. 5. Floor and floor space. 6. Foundation. 7. Walls and ventilation. 8. Windows. 9. Nests and roosts. 10. Dust bath. 11. Sanitation.

1. Location.—A house should be built level, whether conforming with the slope of land or not, because a sloping house makes bad ventilation. The house should face south, be located on a south slope and be protected on the north by trees, buildings or some suitable windbreak. This allows all the sunlight possible to obtain in a day and prevents a northern exposure.

2. Size.—This depends on the size of flock kept. The constant tendency is towards larger flocks, hence larger houses. Some poultrymen keep 500 hens in a healthy condition in one house. But the average seems to be 40 or 50 laying hens per flock. Different breeds need a different amount of floor space, but a good average is 4 or 5 square feet per hen. Thus a house 15 by 15 would accommodate 50 hens. If a greater number of hens are to be kept, a long house all under one roof is the most economical. Two houses under one roof

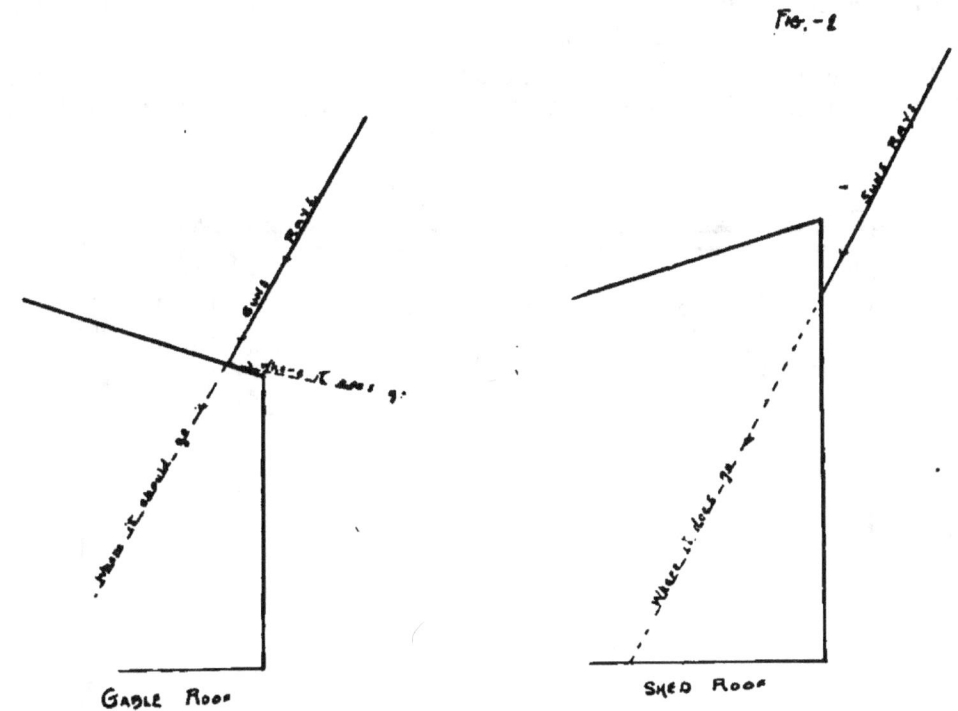

More Sunshine Gets Into a Shed Roof House Than a Gable Roof With Same Floor Space.

saves the expense of one side, etc., and the labor is reduced. The nearer square it can be built, the cheaper it will be in the end.

3. Height of Roof.—Five feet in the rear is about as low as a man can comfortably work in without bumping his head. It should be high enough in front to command a good slope. Make the roof as low as possible so that the hens will not have to heat up any more air space than is absolutely necessary. In building cow stables, the rule is 1 cubic foot of air space per pound of live weight. To apply this to poultry houses would make them impracticable, but we should conform to that rule as nearly as possible.

4. Style of Roof.—The shed roof type is the best for cheapness, ventilation, etc. Light penetrates better in a shed roof front, and what dead, foul air accumulates may be easily let out in front. Gable roofs demand a cupola for ventilation or else a straw loft. The shed roof is the simplest to build, easiest to cut, turns all water to the rear and building paper will last longer on it. More sunlight will strike the floor of a house when a shed roof is used than with any other type. Roofing paper makes by far the best roofing material, because it is the warmest.

5. Floors.—The floor space has been discussed, so the material of the floor should be considered. Dirt is extremely unsanitary, gets full of rats, and is dirty and damp. A board floor is too expensive for its life and is not as good as cement. Cement, while expensive, is rat-proof, dry if well put in and easy to clean. It demands a straw litter to protect the hens' feet, but then all floors should have a litter on them.

6. Foundation.—This should be of concrete and so made as to hold the house 6 inches above the level of the outside ground. It

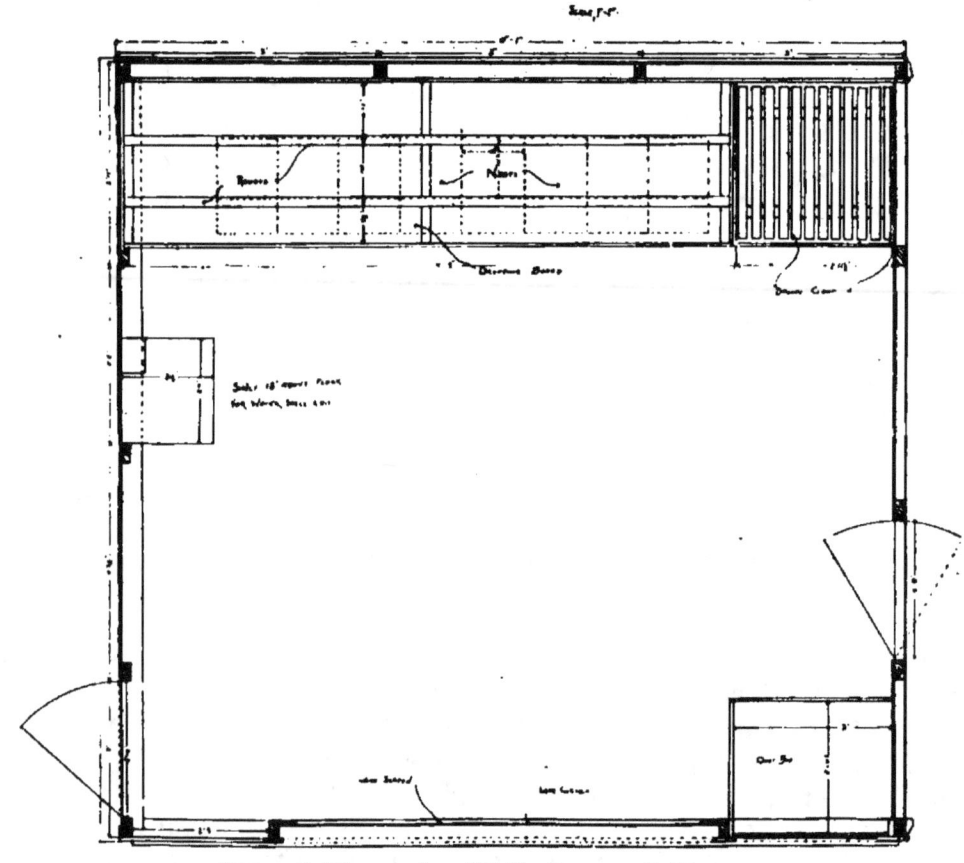

Ground Floor of a Well Arranged House.

need not go more than 8 inches into the ground, just deep enough to prevent frost heaving up the ground underneath the floor.

7. Walls.—There are three things to consider in putting up the walls—temperature, moisture and purity of air. The most economical way of handling these is by putting one thickness of matched pine lumber on the north, east and west sides, and covering with roofing paper. This makes the three sides airtight. The boards should be ship-lap. If the sides and backs are tight and the front open, no draughts can be made through the house, because it is like blowing into a bottle—it can't be done. But the air may work gradually in and out, purifying and changing, but not suddenly.

From the dropping boards, up the back wall and overhead of where the fowls roost, should run an inside wall of matched lumber. In the outside back wall just below the roof should be a door, about 3 feet long and 12 inches wide, which can be opened in the summer. In the front overhead should be another similar opening, to be used at the same time. Thus in the summer an excellent ventilation may be carried on and no draught affect the fowls. This isn't always necessary, but is an excellent thing if the money is at hand to be used.

The walls need not be double, making a dead air space. They are too costly and do not really answer the purpose they are in-

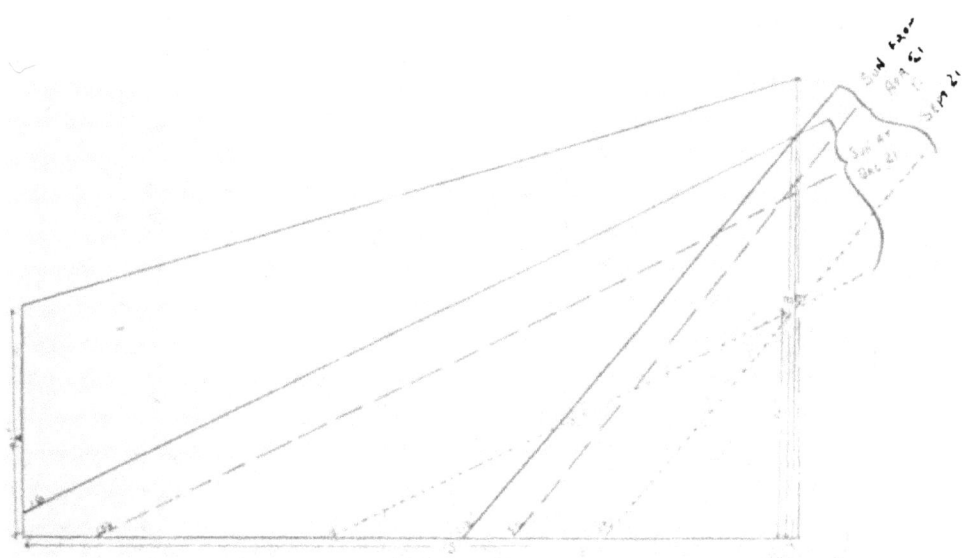

High Windows Allow Sunshine to Reach the Roosts in Cold Weather

tended for. In winter the dead air space would become cold and cool the inner layer of boards, thus causing condensation of the moisture from the warm air inside.

Dampness is caused by the warm, moist air coming in contact with a cold surface, thus condensing the moisture. The warmer the house, the more moisture it will hold in suspension. Keep the house cool, yet prevent draughts, and the dampness problem is reduced. The moisture in a house comes from the hens' breath, drinking pans and droppings.

A house with one open side and all the rest tight, may allow this moisture-ladened air to escape gradually, without condensation, thus keeping the house dry.

8. Windows.—A front should be mostly windows. In order to allow this more or less free circulation of air, all glass windows are out of the question. Glass allows a quick cooling of the inside air and probably causes more dampness than anything else. All wire allows too much free passage of air, so we must use muslin or canvas. Muslin allows the slow entrance of air, prevents draught, allows some light, but not enough. Sunlight is the best germicide known and brings a great deal of cheer in a house, so some glass must be allowed.

The glass window, allowing 1 square foot to 16 square feet of floor space, should be placed perpendicular to the floor, because it allows sunlight to strike both the back and front of the house sometime in the day.

The cloth windows, allowing 2 to 2½ times as much cloth per square foot of floor space as glass, should be placed parallel with the floor and in the upper half of the front. This prevents draught directly on the fowls when open.

9. Nests and Roosts.—The roosts should be at the back of the house under the hood described. At the end of the roosts should be a broody coop, with a slatted bottom, for extra males or broody hens. From the ceiling swing a muslin door so that it will cover the roosts at night. Leave a narrow slit near the top of this door,

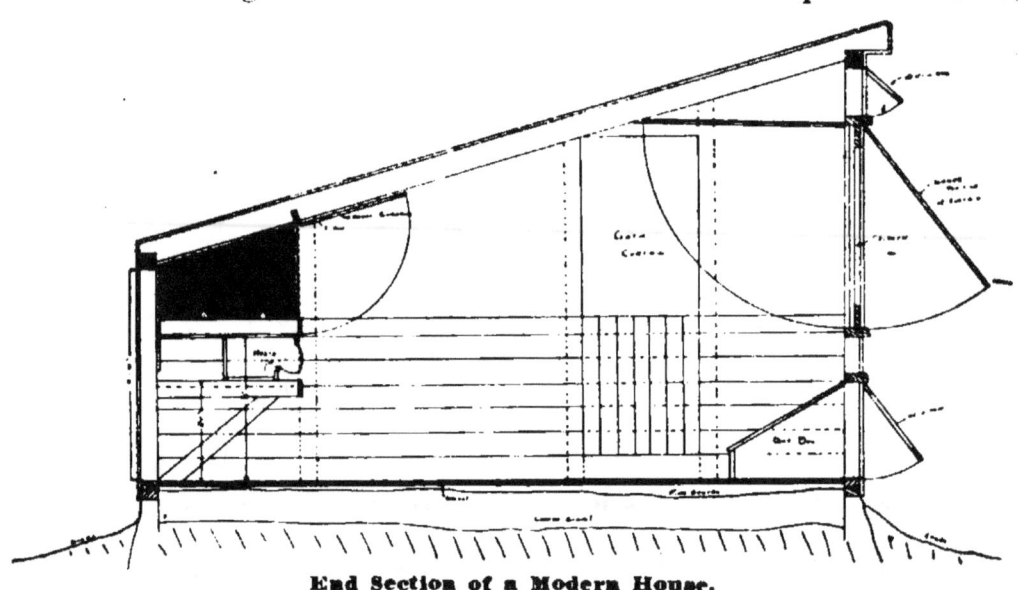

End Section of a Modern House.

so that air circulation may continue and prevent the fowls from becoming too warm.

Underneath the roosts should be the nests, high enough off of the floor to allow freedom of scratching.

10. Dust Bath.—Lice is one of the pests of poultry. The best way to fight this is to allow the fowls the freedom of a dust wallow. So, in front of the glass window where the light will penetrate freely, build a pit and cover with a movable hood. This hood keeps the house almost free of the dust.

11. Sanitation.—This is probably the most important point. Especially where fowls are confined do we find this so. We cannot allow dust that is laden with vermin to blow freely through the houses nor allow droppings to accumulate until foul, if we expect healthy chickens. Now, some may say they allow such things and their fowls are healthy, but we find the savage human race living the same way, without our caring for such a life. An ounce of prevention is worth a pound of cure, and though some things can live in filth, we never know which ones they will be. A house with pure air and a wholesome smell is much more conducive to good health than the opposite.

To best describe the above points, accompanying cuts are shown with this. Maybe you can build a house to suit your circumstances a little differently, but this is a reasonable house that can be built cheaply. It does not pay to put up an expensive house, but a good house will prove advantageous.—A. G. Phillips.

A TOWN LOT PLANT.

Many poultry raisers are situated on small lots. In recent years a strong appeal has been made to such by several so-called methods of poultry raising. Some of these methods contain good suggestions and others contain suggestions that are impractical and unsanitary. If you are raising poultry in small quarters, the first lesson for you to get into your mind is that extraordinary care is required where poultry is kept in close confinement.

Absolute cleanliness is an absolute necessity.

Clean your houses, brooders, runs, etc., every day if your quarters are small. Spade up the runs carefully if they have no grass growing on them. Fight lice and mites all the time by painting roosts, brood coops, etc., with the disinfectants given in this book, and spray houses, etc., with the remedies herein recommended at least every other week in the spring and summer. Feed the feeds recommended, keeping the stock busy, except when you want to fatten it, by scratching in litter. By following these simple instructions you can succeed with your poultry, even if in very small quarters.

Build your houses for large stock, according to our plans given elsewhere and furnish them with as much run as you can conveniently spare. If you cannot give them runs more than 10 feet long and 5 feet wide, do so, but dig up the runs weekly and keep plenty of fresh litter scattered all over, as well as in the house. Arrange your brood coops all together, counting each coop 5 feet long and 3 feet wide. Say you want a place for five broods, then build your five coops together 25 feet long. You can build runs out from these coops from 5 to 10 feet long and with each run a width of 5 feet you will have 25 to 50 feet for your chicks to run on. Build the runs of 1-inch mesh wire and 5 feet high. A brood coop 3x5, with a fireless brooder in it, as recommended in another chapter, will hold from 30 to 40 chicks very nicely in moderate weather, and they can be kept here until 3 months old. If you increase the number in each coop beyond that, you increase the danger to health and the death rate among the chicks. These brood coops should be 1½ feet high at the side away from the run and 2½ feet high on the side facing the run. No limit is put to the number that can be built along together. Three 6-inch boards make the right height for the low side and five for the other. The length of the boards should be according to the number of coops you wish. Thus, for five, 25 feet long, two 12½ feet each would be all right. For only four, 20 feet, use 10-foot lengths. These sides should be of tongued and grooved stuff, not less than ¾ inch thick. The two outside ends should be of the same stuff, but the partitions of lighter material, or even roofing paper nailed to cross-piece of 2x3 stuff at the top and fastened to sides and bottom by a cleat of 1x3 stuff. The coops should each have a cover 4½x5, made of 1-inch boarding, covered with roofing, and it should be hinged with two small hinges at the high side, so as to raise up from the low side. Let this cover extend over the front or high side 12 to 14 inches and 6 inches at the low side or rear. By raising this cover up and hooking with a small hook and eye, at the fence posts that holds the fencing for run, this cover is made to serve as one end of the fence in sunny weather. This lets sunshine into your brood coop, which is very necessary. When the solid cover is raised, cover the coop with a frame covered with 1-inch mesh wire. This frame should be so it will go inside the coop and rest on cleats. It is not fastened, but lifts out. It

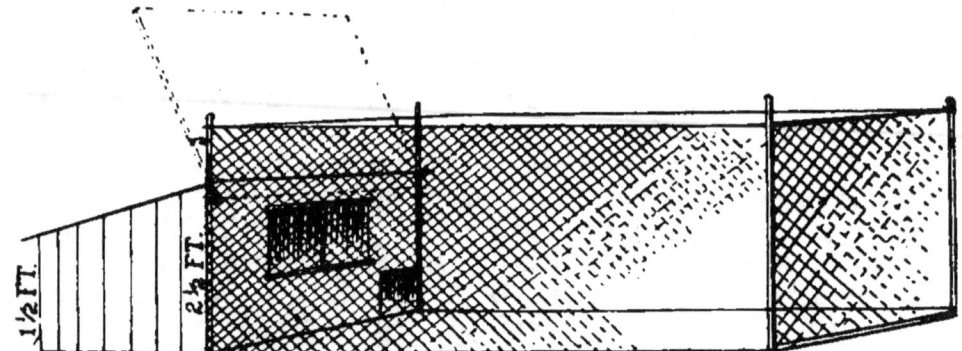

Brood Coop 3x5 Feet and Wire-Enclosed Run 5x10 Feet for a Town Lot Plant.

should be of 1x2 lumber, with two cross-pieces the short way to brace it. When the weather is unfavorable this wire frame is taken out and the solid cover let down and the wire frame hooked up to make the end of the fence. Thus you have the wire frame and solid cover both in use all the time. In placing the fireless brooder, described in another place, in this brood coop, it should be placed within 4 inches of one end, so as to give scratching room at the other end in bad weather. Cut out a place, 6 inches wide and 2 feet long, about 6 inches from the top on the high side, cutting it away from the end where the hover is placed. Cover this window on the outside with heavy muslin that has been soaked in linseed oil, and tack up while damp with the oil. This will give light, and at the same time it will let in a fair amount of air and very little rain. A door for the chicks to go in and out should be cut under the long window, about a foot from the end. The object in having door and window at one end is to keep any drafts or cold air off the end the chicks roost or hover in. Two or three large auger holes should be bored in the low side opposite the door and about 3 inches below the top. These may be left open all the time except in extra cold weather, when rags, straw or something of the kind should be stuffed in them. In real warm weather the top of the solid cover may be raised up several inches at night and the wire frame put over the coop to prevent chicks escaping in the morning. They will not be likely, even if of some size, to fly over the front end before you get up in the morning and raise the "fence" —one of the tops. This brood coop may be raised off the ground a few inches—a 2x4 set edgewise and nailed to the coop along the entire bottom of the high side.

Hens may be used in this plan to brood the chicks.

If hens are used, a larger door will have to be made for them to go in and out. A small door of coarse mesh wire may be put on

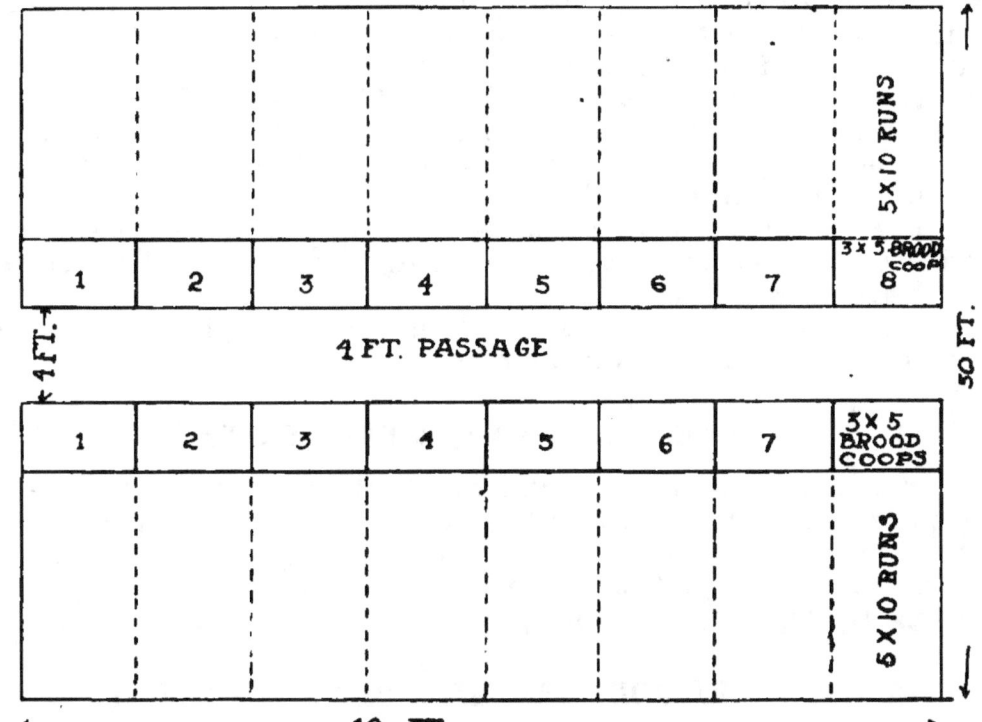

Plans of Arranging Brood Coops and Runs for 500 Chicks in a Space 30x40 Feet.

to confine the hen and allow chicks to go in runs, or even in the outside yard.

By lining this coop with tar paper and using the hover recommended elsewhere, chicks may be raised, even in cold weather, out of doors.

These coops are cleaned by raising up both the solid and wire coverings, and the whole coop is thus exposed and easily reached without much effort. If you have occasion to get in the runs, you can go over the solid cover, using care not to break the roofing paper on the cover. It will not be necessary to get in the runs more than once a week in order to spade them up, clean out rubbish, etc. Fresh scratching material can be thrown in the runs from outside. Considerable expense of building gates is saved by this method. It is better, however, to make gates to go in each run.

Feed hoppers and fountains are kept inside the brood coop. In real hot weather shade may be provided by stretching a cloth along the lower end of the run. This cloth should be 3 or 4 feet wide, and should be off the ground several feet, so that air can circulate under it. If small trees can be inclosed in the runs, it is all the better. Sunflowers may be grown in the yard, planting them very early in the spring and keeping them inclosed with a small wire fence until they are large enough so the chicks will not damage them. A place 30x40 feet will hold two tiers of brood coops of eight each, with yards 5x10 feet. This will make sixteen coops and, estimating a little over 30 to the coop, it will take 500 chicks on a space this size. The brood coops and runs should be arranged so that two sets of them will have their coops at the same end and thus by going down the alley, care may be given to 500 chicks in a distance of only 40 feet. We give an illustration of an arrangement of brood coops for 500 chicks. Of course, more than eight can be built together if you have a space more than 40 feet wide that you wish to devote to brooding chicks.

Great care must be taken to keep absolute cleanliness, plenty of pure food, as well as spraying frequently for mites and watching the chicks every week for lice.

Remember, again, the greater the number in a given space, the more the trouble and closer the care.

HOW TO GROW LARGE COCKERELS.

Just as soon as the cockerels can easily be distinguished from the pullets, put them in separate runs, at about 4 to 6 weeks, and feed them a good growing ration. By keeping them separate all the time from the pullets they will grow much more rapidly and increase to larger size than if allowed to run with the females.

HOW TO GET MORE PULLETS THAN COCKERELS.

Keep your hens as quiet as possible.

It is claimed that mating an old cock bird, 2 or 3 years old, to 15 or more vigorous pullets, will produce more pullets than cockerels. The eggs will not be so fertile as a smaller mating. Eggs hatched late in the season will produce more pullets than those hatched earlier, it is said.

The feed should be very high in nitrogenous feed, like wheat, meat, linseed meal, green bone, meat scraps, etc. Feed very little corn.

A SIMPLE FEED HOPPER.

A simple feed hopper is here illustrated and may be used to advantage for growing stock or breeding stock. It will be noticed that one-half of the top is hinged so it can be raised and feed put in or dirt cleaned out. The hopper here illustrated is 8 feet long and is 8 inches wide. The ends are solid. The side board is 4 inches wide. A strip projecting ½ inch is nailed on top of the side boards to prevent the chicks pulling out feed. The upright slats are 3 inches apart and common lathing is used for making them. The

A Simple Hopper That Fowls Can Eat From Out of Both Sides.

hopper may be set on a 2x4, 2 feet long, one at each end, nailed across the hopper to prevent it being turned over. The height between the bottom part and top of the hopper, or the space they reach in and get feed, should be from 6 to 8 inches.

HANDY POULTRY CATCHING HOOK.

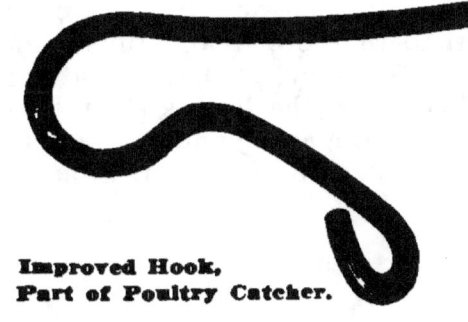

Improved Hook, Part of Poultry Catcher.

There is no use calling the dog, the children and several neighbors when you want to catch a chicken, and then having a big hurrah, race and scare for all the poultry on the place. By using a catching hook you can slip the hook over the leg of the fowl you want to catch, draw it gently towards you, and create no disturbance among the other members of the flock. At the same time you will save much useless worry to yourself. It is well known among poultry breeders that fright decreases the egg yield and should be avoided if possible.

The hook is made of a 6-foot piece of No. 10 steel wire and a broom handle. The hook is so bent that it is larger where the shank of the fowl is held than at the opening. Thus the fowl is not injured when caught and cannot easily get away. The wire is less conspicuous than the wooden end which attracts the fowl's attention when you slip the hook over the shank. The wire may be braced by wrapping it with No. 8 steel wire for the last two feet next to the handle and fastening this wrapping to the handle. At least two feet of the wire next to the hook should be unwrapped wire. You will note that the point of the hook is bent back, so as to prevent injury to the fowl.

HOME-BUILT HOPPER.

The feed hopper here illustrated may be built of old dry goods boxes, and a hammer, saw and square are all that are necessary to do the mechanical work. Cut out two boards for the sloping sides, making them 4 inches wide at the bottom, 10 inches wide at the top and 20 inches high. These are triangular in front, as shown in the illustration. A board 15x18 inches makes the front and one 15x20 the back. Two boards 5x8 inches tacked on at the bottom keep the grain in at the sides. A board 15x5 inches closes the front and one 15x2 inches should be tacked on top of the front and two 5x8-inch end boards to prevent the chicks pulling out the grain. Care should be taken that the angle of the two sides is not too sharp, so that the hopper will be topheavy and tumble over if filled with grain. The front upright board should lack 2 or 3 inches of going to the bottom of the hopper. The bottom of the hopper should be 15x12 inches. This will tend to make it stable on its feet. Make the top some wider and longer than the hopper and thus keep the feed dry. Hinge top at the back, but the hinges should be under the cover and not on top, as shown in the illustration.

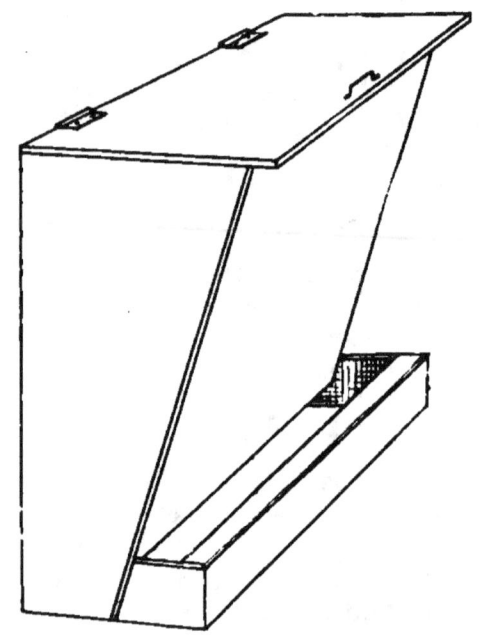
Popular Type of Feed Hopper—Home Built.

Knowing and Breeding the Heavy Layers

METHODS OF SELECTING THE LAYING HEN.

We treat fully in another chapter of the pelvic bone test and the methods given here are also good to use in connection with this test. In fact, it is only by an application of all the known methods that the best layers can be arrived at without trap nesting, as some of the best layers will be found to not have all of the specifications of any method.

One of the methods of increasing the productiveness of your flock is to select only the very early layers, and breed from them. The first pullets that lay can usually be told very easily by the small eggs, their reddened comb, frequent singing, etc. In all flocks there are a few pullets that lay considerably ahead of all others. These will be found to lay a far larger number of eggs during a year than their sisters that lay later on.

Another method, by trap nesting for two months in the laying season, a very fair idea of the laying ability of each hen can be determined. Some hens lay every day, some every other day, and some every third day, while some scarcely at all. After laying out a clutch of eggs, some hens rest three, four, five or six days, while some rest as much as two weeks. It is usually found that the hen that lays right along each day will only rest a few days until she begins to lay again. This hen is a valuable hen to keep and breed from, and not the hen that lays every other day or the one that rests two or three weeks between laying spells.

In selecting layers, three things should be carefully considered.
First. The shape.
Second. The color of comb, wattles, face and plumage.
Third. The actions of the hen.

Shape.—It cannot be said positively that the egg type of the hen is yet known. The majority select a hen that is wedge shaped, wide behind, narrower in front, and low down, with legs not too long or too short, but wide apart. The back is usually longer in the best layers. This shape here indicated gives big lung capacity, big organs for digesting, and large egg bag. The crops of layers will most usually be found full at night, while those not laying are usually only partially full. The fingers may be greased and inserted in the vent and hens laying heavily will be found to have large egg bags, and small layers small egg bags.

Color of Comb, Face, Wattles, Etc.—In the laying hen, face and comb are nearly always a bright red, especially in pullets that are just beginning to lay. A pullet that is soon going to lay will show very red comb, etc. The plumage also of birds that are laying is usually brighter about the time they begin to lay than any other time. Heavy layers in the buff and some parti-colored varieties can be told by a dingy effect that their feathers have after a season's laying. A hen that has laid heavy all season has a much more faded plumage than one that has laid little. It can be safely stated that a heavy laying hen always has bright comb, face, etc., unless she is sick, and that shows by other signs, also.

Actions of the Hen.—Early laying is one of the best indications of a good layer. A hen that is going to lay will usually be found singing and "chuckling" to herself a good part of the time. A laying hen is also more active, more inclined to be hunting and digging for something to eat and is more nervous than the non-layer, who is usually drone-like in action. The real fat hen, as well as the real lean hen, is not a good layer. You want a happy medium and this gives a fair amount of activity. Many heavy layers can be told by a duck-like walk that is caused by breadth of body, a fair amount of fat and width between the legs.

Other Things to Note.—Neither the overly large or the undersized hen should be selected for high egg production. Of course, in considering size, you should consider the standard weight of the breed. For instance, a 5-pound Wyandotte hen would be more likely to prove a good layer than a 5-pound Brahma, while an 8-pound Brahma would be more likely to prove a good layer than an 8-pound Wyandotte.

If you are improving your egg laying or working to that end, it is just as important to have cockerels that are from egg laying mothers and grandmothers as it is to select the good egg layers to mate them to.

The Belgian breeders say to select the cockerels that crow first and mate them to the pullets that lay first, and thus you will increase the vigor and egg laying qualities of your flock. A surer method is to select the first cockerels that crow, provided that they are right in other qualifications, and then selecting from these early crowing those whose mothers and grandmothers were known to have a high egg yield—early layers, good shape, and tested by the pelvic bone test—or, better, tried out in the trap nest.

Some breeders select winter layers from the yearling hens in September and October by taking only those in heavy molt. A hen that does not begin to molt heavily in these months will be so late molting that she will not prove a heavy layer that winter. Late hatched fall pullets are an exception, but are so rare as not to count in large flocks. By this method, take the hens that are full of pin feathers in October. By December 1 such a hen will be through her molt and soon begin to laying.

THE PELVIC BONE TEST.

This test has been largely used and much claimed for it. It can be said to be a sure indication of the hen that is then laying, but to take it and tell how many eggs a hen will lay during the year is a different matter. However, by practicing the system on the hens every two weeks, the poor layers can soon be weeded out.

The skin in most animals when they are in good condition is soft and pliable to the touch. In hens this elasticity is most noticeable in the hinder part of the body, from the legs to the vent, and especially when they are in full laying. If a bird has been out of condition for any length of time, or even when overfat, this pliability of the skin is then apparent. There is also a corresponding expansion and contraction of the pubic bones of the pelvis noticeable at this part of the body, according to the laying or non-laying condition of the fowl, and on this fact is based the so-called discovery of these several systems. When it is understood that laying almost wholly depends on condition, the fallacy of any such system of picking heavy layers is apparent.

No matter what width the pelvic bones may be apart, if a hen is overfat she will not be a heavy layer; and any reduction in flesh below normal laying condition that has a tendency to interfere with the vitality of the bird will both check her laying and cause a speedy contraction of the abdominal parts mentioned. However, when once a correct knowledge of the application has been acquired, this combination of fair condition, wide pelvic bones, and pliability of the skin beneath the fluff will prove a sure guide as to whether the bird handled is layng at that particular time. If hens are carefully looked over at stated intervals and those found to be not laying are removed, the drones of the flock will soon disappear, for they cannot possibly escape detection.

When lifting a hen to examine her condition, the most handy way to take hold of her is by placing the hand on her "shoulders" from the front; and slipping the thumb under one wing, and the fingers under the other, to grasp them at the butts firmly but gently; the hen then can be handled without any fuss. Then, by placing he other hand on the fluff or rear part of the bird's body from the underneath side, or by slightly turning her over, with the tips of the fingers, the two pelvic bones which lie one on each side below and adjacent to the vent, can be located. Almost invariably in a low conditioned bird the points of these bones stand clearly defined to the touch. As the bird makes flesh they gradually become covered, till, in an overfat bird, there is some slight difficulty in exactly locating them on account of the thickness of the overlying tissue. The structure of fowls varies greatly in its formation at this part, some birds having a much larger space between the pelvic bones than others, even as chickens; and others, when matured, having the points of these bones not more than one-half inch apart; others, again, will have them at various widths, even up to two inches. To a practical poultryman who understands feeding for condition, and who consequently is well versed in the handling of fowls, the position of these bones, when taken with the bird's general condition, is a tolerably good guide as to whether she is laying or not. He can also determine how long it would take to put a particular bird into laying condition if necessary. When, from any cause whatever, the bird is not laying, there is almost immediately a drawing together and tightening of the skin, and a closing towards each other of the points of the pelvic bones. There is also a corresponding relaxation in those parts as laying approaches. As size, age and breed all to some extent control the

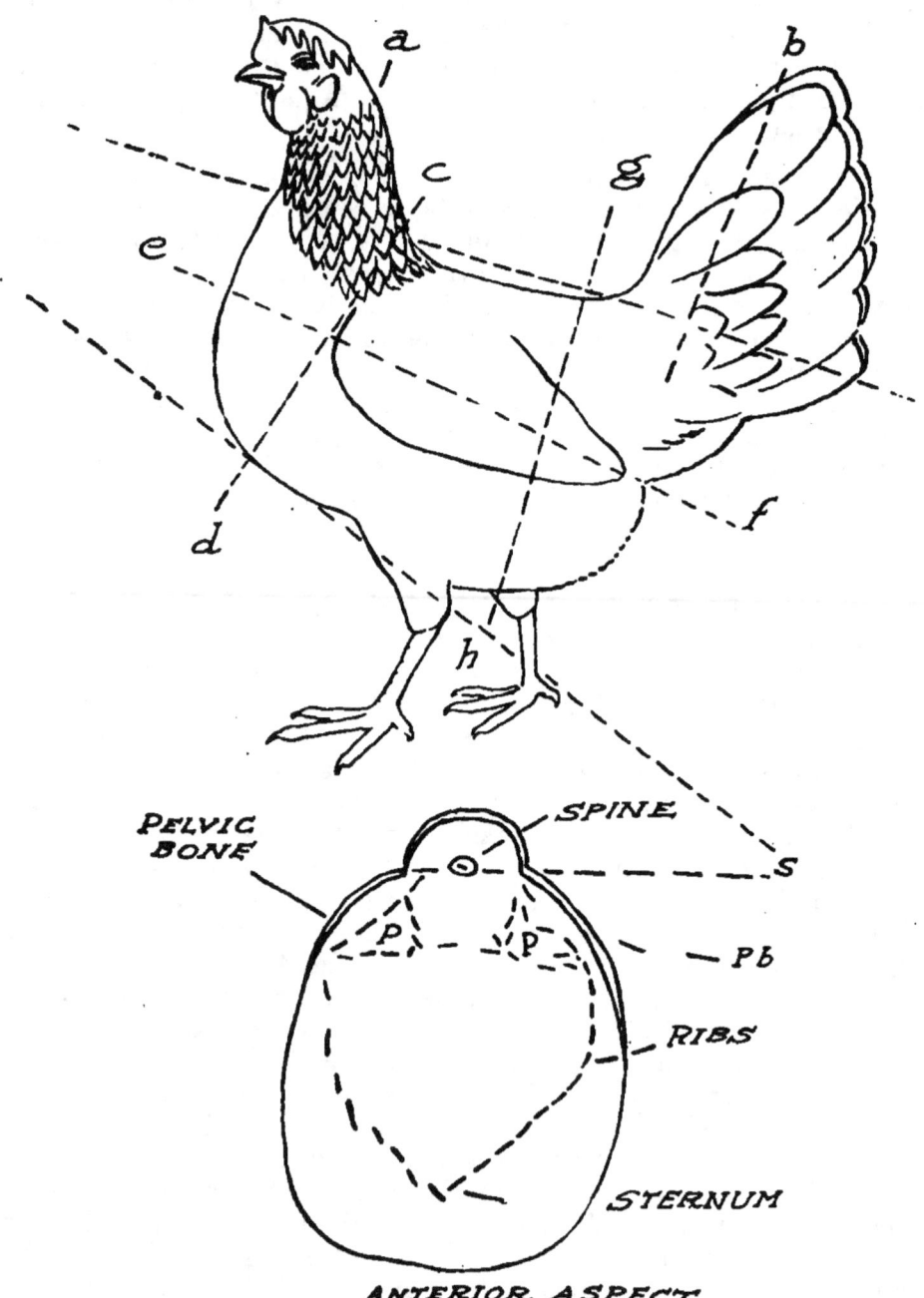

ANTERIOR ASPECT
Diagram for Use in Selecting the Laying Hen.

structure of each bird, the actual widths between the pelvic bones cannot by itself be taken as positively demonstrating the laying condition of any hen. Speaking generally, the bird that is fairly wide between these bones may be looked upon as a good layer; just as one that is wide and deep in the fluff when viewed from the rear is likewise usually considered. The condition of the bird can only be ascertained by handling; and without the knowledge of her condition, as drawn from the actual touch, no external appearance of a hen will correctly indicate whether she is laying or not.

Fowls generally molt during the late summer or autumn months, and while they are growing their fresh coat of feathers egg production usually ceases. Molting occupies from six to eight weeks; and, unless she is hatching or raising chickens, a hen should be engaged in laying throughout the rest of the year. Many hens will take an occasional rest for a week or so after a long spell of laying; and, unless those rests are prolonged to the possible detriment of the total egg yield, no notice need be taken of them; but the hen that does not do anything towards paying her board during any four consecutive weeks—excepting when she is molting—should be removed from the flock and either fed to promote laying or be sold for what she will bring.

EGG-TYPE FORMATION.

The following are the requirements which each hen must fulfill: (1) She must be healthy. (2) Comb red and full. (3) Eye bright and lustrous. (4) Neck not short. (5) Breast broad and sloping upward, long. (6) Back long and broad. (7) Abdomen wide and deeper than breast. (8) Well spread and rather long shanks. (9) V-shaped three ways, viz., (a) on sides, rear to front, (b) top and bottom, rear to front, (c) base of tail, from abdomen upwards. (10) Well spread tail.

HOW TO STUDY EGG TYPE.

Select (a) three hens which in your judgment show the best egg-type conformation, (b) three that show the poorest egg-type conformation.

For each of the six fowls selected record the following: Length of back from origin of wing (a) to base of rump (b); diameter of body, point to keel (d) to base of neck (e); diameter of body, rear of keel (b) over back at (g); circumference at (c-d); circumference at (g-h); circumference at (e-f); distance across pelvic opening (p-p); distance across pelvic opening from spine (s) to point of pelvic bones (p-d); vertical distance across from (p-b) to rear of sternum.

HOW TO BUILD A CHEAP TRAP NEST.

A trap nest, to be successful, must not only catch every hen that enters, but must prevent the entrance of other hens. The one here shown does that to perfection and is very simple to understand as well as to construct. In constructing one of these trap nests the gate is made of thin box boards, cleated top and bottom, and should be about 12x18 inches. A hole is sawed in the lower half 8 inches square. A small gate is made by connecting two pieces of wood together with three wires, 8 inches long. Holes are bored in each end of these pieces of wood, so the gate may slide up and down on two upright wires that are fastened to the gate cleats. This gate is held up by a wire trigger, having a shoulder bent on one end, on which the gate rests when open. To close door shows that the same kind of a wire shoulder is used to fasten the door, which is done by simply pushing it shut. To open the door this latch is raised with the thumb until the door is released.

These doors may be fitted to any kind of a box or barrel, but we find it best to nail a 1x4 up edgewise across the box, about 4

View of Trap Nests—The Left-Hand One Is Closed and the One on the Right Opened to Show Triggers.

inches from the front, to keep the nesting material from interfering with the gate. This also makes the hen more certain to trip the trigger in entering the nest.

The partitions of the nest are made mostly of cheap muslin, and extend from the roost platform to the roof, so that the hens cannot roost on them. The fronts are made in one section and held in place by three nails, partly driven, so that they may be readily moved should the interior of the nest need spraying or whitewashing.

The cost of making one of these doors with the gate and trigger attached is about 25 cents each.—W. A. Lamb.

A ONE-WAY TRAP.

The device or gate given below can be used as a trap to a nest, or it can be used in the system of trap nesting to secure fertile eggs described under the head of "Incubation and Brooding," which see.

This invention was made by the veteran, A. I. Root, editor of "Gleanings in Bee Culture."

Take two white pine boards that you can readily whittle out the curves as shown in the illustration. These pieces may be 10 inches long and 2½ or 3 wide. Then you want two sticks, about an inch square and 8 or 9 inches long. Nail them together so that the width of the circle where the hens pass through will be 6 or 7 inches. This space of 6 inches wide will let a Leghorn hen through all right. The swinging wires should be tinned and large enough so that the hen cannot bend or spring them easily when she wants to go through the opposite way. These wires are bent L-shaped, or a little more than square at the corners. The long arms should be curved a little where the hen goes through, so as to indicate where they are to push between the two wires. Drive a couple of staples about 2 inches apart in the upper stick. If these wires moved out squarely, like the lid of a box, a hen could get through without much trouble either way. But we want them so as to spread apart as shown at the dotted line. To do this, take an ordinary twist drill and drill diagonally into the upper stick. Push the short end of the wires into these drill-holes, and, if drilled rightly, as each wire is raised it swings off out of the hen's way, and allows her to go through easily. After she passes out, the wires drop back, just clearing the nails partly driven into the bottom stick. When the fowl attempts to get back through the trap the way she came in, the lower ends of the wires spring under the nail heads as you see.

HOW TO KEEP EGGS FROM HATCHING.

The popular superstition is that a small drop of grease will keep an egg from hatching, but this is not true, as it requires covering the grease over considerable part of the egg in order to kill the germ. Any kind of grease getting on the outside of the egg during incubation will more or less affect the hatching of the egg by smothering the germ. If you put enough grease on eggs to kill the germ it will be enough to show and be easily detected. One of the favorite ways of preventing eggs from hatching, if you have eggs that you are selling for market purposes and the eggs are from mated stock, is to put a single pin prick in the broad end of the egg. This will nearly always prevent hatching, and is not easily detected, unless especially looked for. Another method is to drop the eggs for a few seconds in boiling hot water. They should be put in a wire frame and let in a few at a time. Of course, no one with any self-respect would so treat eggs and then sell for hatching. All eggs that you expect to sell on the market should have the male birds removed from the females at least two weeks before the eggs are offered for sale. Eggs that are not fertilized keep better and the hens will produce a larger output, being free from worry of the male bird.

LINE BREEDING.

Line breeding is the best method you can follow to start a strain. It may be described as selecting a male and female that are as near

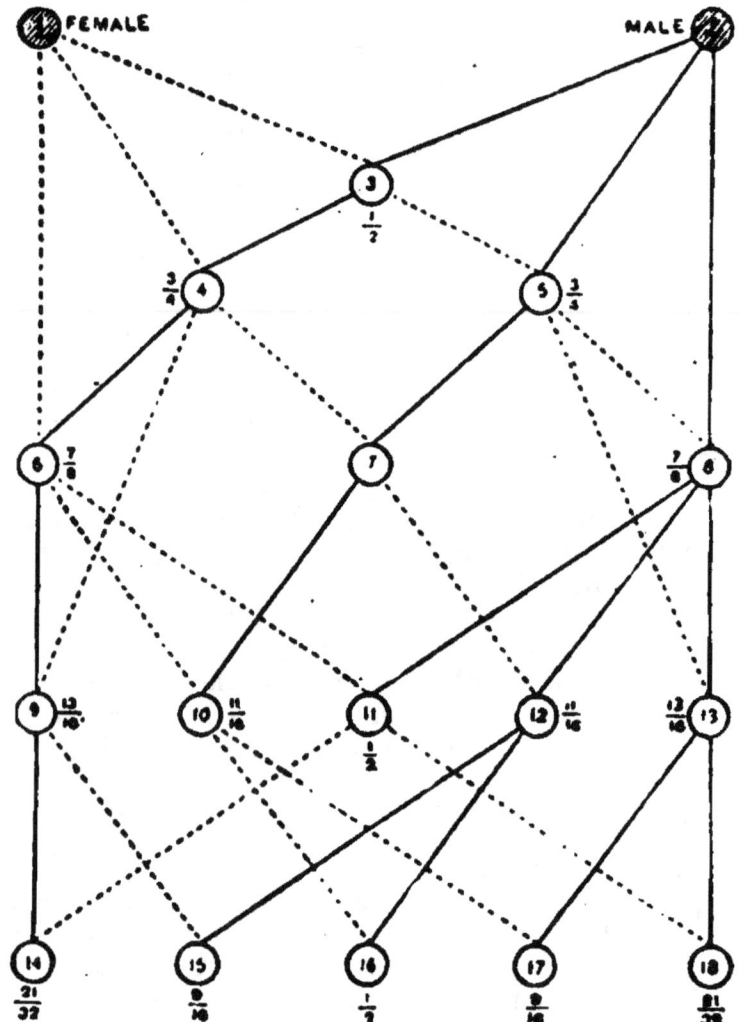

Felch's Chart for Line Breeding to English Strains.

ideal as you can get hold of and mating them. The following season mate the sire to the best one of his daughters and mating the original dam to one of the picked cockerels of the first year's mating. The third season you breed together two birds of the second season's mating, one from each side, and you thus have 50 per cent of the original blood of both birds on each side and still do not mate brothers and sisters. The chart given on line breeding herewith is known as Felch's chart and was originated by that veteran poultryman, I. K. Felch. He is also due credit for much work in developing the line-breeding system.

In this chart each dotted line represents the female, having been selected from the group above, while the solid line represents the male so selected. Each circle represents the progeny. Female 1 mated with male 2 produces group 3. Female from group 3 mated back to male 2 produces group 5, and has three-fourths the blood of No. 2 and one-fourth the blood of No. 1. A male from group 3 mated to female 1, the dam, produces group 4, only one-fourth of the sire blood, No. 2, and three-fourths of the dam blood, No. 1. A male from group 5 and a female from group 4 produces group 7, one-half the blood of the orignal birds for the second time. The original male is now mated to females from group 5, and this produces group 8, which has seven-eighths of the original male blood, and a cockerel mated back to the original female, No. 1, produces seven-eights of the original female blood. A male from group 8 and female from group 6 produces group 11, which are the third set of birds to be half the blood of the original pair. By mating a male from 6 and female from 4, group 9 is produced, and this is thirteen-sixteenths of the original sire blood and the same of the original female. A female from this group, 11, and a male from group 9, produces group 14, your new strain on the female side, or twenty-one-thirty-seconds the original dam blood. A female from this same group and a male from 13, produces your new male strain, or twenty-one-thirty-seconds of the original male blood.

This system is kept up on and on and no new blood is introduced. It is rather complicated to understand until practiced, and then it becomes easy. This method is not inbreding.

SOME "SYSTEMS."

The Philo.—The essential point of this system is keeping fowls in small coops or houses. The fireless brooder is used and the chicks are put in one of these houses when hatched and kept all their lives confined in them. The coops are 3x6 feet and about 3½ feet high, and six grown fowls, or 25 or 30 chicks, are kept in each one. The ground where they are set is made higher than surrounding ground and carefully dug up each day. Feed is put in litter to make the fowls exercise. One end of the coop, about one-third, is floored about 2 feet off the ground, and this is used for a dropping board and for nests. The dropping boards are taken out and cleaned daily. The roof is made of wood boards, covered with roofing and hinged the long way, so that it can be raised up. Under this roof a frame fits down inside the box, resting so as to slide out at the ends. It is covered one-half with 1-inch wire netting and one-half with muslin. The roof is left up all the time, even an inch or so in cold weather.

The roof and frame of wire netting and muslin, in the sum

mer, are set up so as to make a "V"-shaped roof, one-half being solid roofing and the other half the muslin and wire covered frame. Triangular end pieces are used to close the ends.

The methods of feeding are very little different from those practiced by all poultrymen.

As the fowls are all the time in this small coop, or house, it requires infinite care in keeping it clean, feeding regularly, spading up the ground, etc. All this requires extra labor, and it is doubtful if it pays on a large plan. Where the range cannot be had, it may prove profitable for only a small flock, to be tended mornings and evenings. This system of close confinement has not been used long enough to tell if it will impair the vitality of the flock, but it is generally believed that this method in a few years will deteriorate the stamina of a flock.

The Corning Plant.—A father and son in New Jersey claim a profit of $6.40 per hen. They estimate the cost of feeding each hen $1.50. The eggs are packed and sold at first-class hotels and restaurants at 10 cents above the market price, with guarantee of at least 40 cents during the year. They keep only pullets in flocks of 1,500 each in houses 16x160 feet. They are the open front patterns, set on posts 5 feet above the ground and the space under the house is the only yard the hens are allowed. In the winter they are kept in the houses. This is the market egg flock, but for breeding purposes regular pens are mated up of 15 yearling hens to one cockerel, and it is claimed this mating breeds more pullets and most fertile eggs. The chicks are all hatched between April and June.

Their method of feeding does not differ materially from that used by other practical breeders.

The profits claimed by these people have been criticised by practical poultrymen because no interest was allowed for the money invested, insurance, or depreciation in value of buildings. Also, the claim of selling a large number of eggs and pullets at fancy prices, without any expense of advertising, is seriously doubted by poultrymen who have spent years in the poultry business and been unable to sell poultry at fancy prices without building up a reputation by advertising, winnings in shows, etc.

The Miller.—A long description of how to build a complicated trap nest is one of the main features. A brooder, to be heated by the animal heat rising from decomposing barnyard manure, is recommended. A pipe is covered with decomposing manure so as to be heated and passed into the brooder. This manure-heated brooder is impractical and was tried many years ago and discarded by practical poultrymen. The methods of feeding recommended are about like those practiced by poultrymen for years before this system appeared.

Incubating and Brooding

SECURING FERTILE EGGS.

The first, and most important thing, is to have vigorous, healthy, well matured, but not too old, breeding stock.

Second. Do not mate too many birds in one mating, from five to 12, according to breed and vitality of the male bird. The Asiatics require smaller mating than any breeds, while the Mediterraneans will allow the largest number.

Third. Feed should be with a view to produce fertility rather than great numbers of eggs. The feed should not contain over 10 per cent meat scraps, beef meal or fresh cut bone. It should have plenty of nitrogenous feed and not over one-third corn. A very good ration is beef scraps, 10 per cent; wheat, one-third; cracked corn, one-third and oats, one-third. Also see recipes for some very successful feeds under proper headings. At least two feedings a day should be fed the fowls in a litter from 12 to 18 inches deep. The litter should be clean and fresh, so that their feed will not be contaminated by being stirred up in it.

Remember, your breeding stock must take plenty of exercise. If it has range it will take enough.

Some breeders practice a system of trap nesting that releases the females after laying into the yard where the male is. At night you will thus know every female that has laid, and that only layers are with the male. The next morning your females are placed in the yard alone and the male in the adjoining yard. This method is said to insure a high degree of fertility. The hen that lays seldom will not be in the yard with the male as much as the one that lays often. The theory is that this is the proper method, as hens not laying much do not require the service of the male to increase fertility. (See "Trap Nests.")

Some breeders practice a method of having two males for each yard, and alternate them once a week. The theory is that a week's time gives a male an opportunity to become better acquainted with the females in the pen. This system cannot be practiced where eggs are wanted from a special male bird. Where this is desired, the best method is to mate the bird to individual hens, having small pens for each hen, with a small roosting or colony coop for each pen. The pens need not be over 5x10 feet. This saves trap nesting, but it is nearly as difficult to care for the hens in this manner as trap nests, but the time is all taken nights and mornings, and not all along throughout the day, as in trap nesting. In this manner the male bird can be put in each pen half a day at a time. Not over four or five hens should be mated to one bird where this individual system is practiced.

TESTING EGGS BEFORE SETTING.

Various methods to tell whether an egg will hatch or not have been suggested. It may be safely stated that it is at present an impossibility to tell whether an egg has the proper germ cell to insure its hatching. But some idea as to its age can be had by simple tests.

An egg 3 to 4 weeks old stands far less chance of hatching than one a week old. Shape and size have something to do with hatching. Very large or extra small eggs usually give very poor hatches. Ill-shaped and deformed eggs also hatch poorly.

A method offered for sale and claimed much for is that the large end of the properly fertilized egg has a rough, cross-like mark on it. It is claimed that this shows the proper male fertilizing element entered the egg. It is a fact that the male germ enters the egg up in the ovary and before the egg has taken on any shell material at all. So this theory has no foundation.

Another method is to use a common egg tester and the small and light colored yolks are infertile and the large, rich colored ones are fertilized. It is true that the richer yolk may cause the chick to grow better after hatching, but the size and color of the yolk have nothing to do with the hatchability of the egg. The yolk is simply feed stored up for the chick's use for the first few days after it hatches until it can rustle for itself. Both the male and female germ cells are located in the white of the egg and all the chick development before hatching is in the white.

The color, as well as the size, of the yolk is affected by the feed. A large, rich, deep yellow yolk indicates the hen has been fed a strong, fat feed, with plenty of green stuff for a few days before laying the egg. The light colored yolks indicate nitrogenous feeds were eaten.

It has been claimed that by the use of a miscroscope the germ cells could be seen in the egg, but as they are practically the same color as the white of the egg, this method is very unreliable.

By breaking an egg, two cord-like attachments can be seen floating in the white, and the germ cells are attached to these, but it is impossible to distinguish the cells through the shell. These cord-like attachments are in both fertilized and unfertilized eggs.

Another method is to look through an egg tester for a large yolk, or spot, which is claimed indicates the germ. This method is unreliable, as the life germ is miscroscopical.

In recent years one of the widely advertised methods has been what is known as the "specific gravity" method. This is simply telling the specific gravity of the egg. It is claimed that eggs low in specific gravity are poor to hatch. As an egg dries out from the time it is laid, eggs low in specific gravity will, of course, be older than those in higher gravity. As the water evaporates out of an egg the air cells at the large end grow larger.

These patented instruments to tell the specific gravity of an egg are no surer than the old method of testing eggs by dropping them in a large glass three-fourths full of water. Common table salt may be stirred in the water until an egg that is known to be fresh will barely lay on the bottom. This will cause the staler eggs to rise off the bottom easier.

About 10 per cent of salt may be used. With this much salt, the fresh eggs sink immediately to the bottom. If 3 or 4 days old, they sink only below the surface, while, when a week old, float partly out of the water.

Real stale eggs will float out of the water. This simple method will give an idea of not only the age, but amount of water in an egg.

Illustration of Simple Method to Test Eggs.

An egg that is fresh and floats with the big end off the bottom of the glass, should not be set, as it is already too low in water—specific gravity—to hatch well.

An instrument has been widely advertised to tell the sex of eggs. This instrument has been demonstrated as unreliable. So far no method of telling the sex of eggs, before they are hatched, has been discovered. The old superstition that eggs of certain shape will hatch cockerels has been shown to be a delusion.

HOW TO MAKE AN EGG TESTER.

Take an empty box used for breakfast oats or any of the prepared cereals. Cut off one end, and then cut away the pasteboard (on one side) on a curve so as to come against your forehead. On the opposite side make a V-shaped opening to fit the nose; in fact, you want to fix this pasteboard box so it will come up tight and close around the eyes, nostrils and forehead. When put up against your face you have a dark box. In the opposite end of the box make, with a small-bladed knife, an oval opening. It should be just about the size and shape of a good-sized spectacle glass. The egg is to be held up against this opening on the outside; but in order to have it fit tight against the pasteboard you want to paste around the opening a circle of dark colored woolen cloth. The idea is to shut out every particle of light, especially any gleam of light that might get in around the egg; then if the light is cut off in the same way when the pasteboard box comes up against your face, you will have a little "dark room." With this instrument you can tell whether a white egg shell has the germ started, generally in 48 hours. In 72 hours, or three days, you can tell very plainly. Point your egg tester right toward the sun; or, if after night, hold the egg near a strong lamp or electric light.—A. I. Root.

SOME HANDY ITEMS IN INCUBATION.

Don't be afraid to use plenty of moisture after the second week.

Don't be afraid to air from 20 to 30 minutes in a room that is comfortable for you to stay in.

If your incubator is running in a cellar you will not need to bother with the moisture problem, as it will largely solve itself.

Whether you have hens or incubator, wet the eggs in water that feels warm to the hand on the nineteenth day.

Be careful that pipped eggs are just dampened in the water, but eggs that are not pipped may be let stay in the water one-half minute or more. This will improve your hatch considerably.

A woolen cloth wrung out of hot water laid over the eggs after the eggs begin to pip is also good when you are using an incubator.

Eggs that have live chicks in them on the twentieth day will show a quivering motion when put in water that is as warm as your hand can stand well. As soon as they begin to quiver remove them and put them under hen or in machine. Another method of discovering a live chick at the time the eggs are due to hatch is to put the eggs on a piece of glass and drop a little ice cold water on them. Frequently they will move without the ice water.

Eggs should be tested where set under hen or in an incubator on the fifth day. Eggs that are perfectly clear on that day and do not show a dark spot may be removed. However, if you are uncertain, leave them until two days longer. Any incubator that you have will have a description and illustration of the development of the chick germ for your guidance in testing. If you have not such a booklet and are using hens, it is useless to test the eggs under the hen, except to remove infertile ones on the fifth or seventh day.

Very little can be done to improve the hen's method of handling the egg except to dampen them weekly and wash off filth that may be on them. Filth should be washed off of setting eggs as soon as found.

While it is generally useless to try to save chicks that do not come out of the shell by the twenty-first day, yet you can frequently save them by carefully pecking around the shell with the point of a knife and carefully dampening the shell, if the chick is alive and cheeps. Such eggs should be removed from under the hen and wrapped in a warm woolen cloth and placed in a warm place. If in an incubator, place a woolen cloth, wrung out of as hot water as you can bear, over the eggs treated in this way.

A NATURAL HEN INCUBATOR.

Where you wish to use hens for incubating eggs, they should not be allowed to set all around through the house, one or two in a place, but they should all be in one house, or if you have an old barn or shed they may be put in there. A room or shed 12 feet each way can be made to accommodate 20 hens.

A "natural hen incubator" should be 18 inches high and each compartment should be 12 inches wide and 4 feet long. To illustrate; for a shed or room 12x12 feet here is a good plan to build one for 20 hens. If your shed is longer, add on extra setting compartments.

Build a box without top or bottom, 18 inches high, 10 feet long and 8 feet wide. Run partitions a foot apart the 8-foot way, making compartments 12 inches by 8 feet. Divide these compartments in two at the center, making them 12 inches by 4 feet. This will give you 20 compartments. At the ends, away from the middle of each compartment, take a board 6 inches wide, 1 inch thick and 10 feet long and nail across the top, tacking it to each partition. This board should be nailed so as to leave an opening 18 inches at the ends away from the center partition. Hinge doors to these boards, having the doors 18 inches by 3 feet, except one, which will have to be 18 inches by 4 feet. Thus each door covers three nests at each end, except one, which covers four. The space in the center, between the two 6-inch boards, may be covered with wire netting, 2-inch mesh, or strips of boards. This makes the top. Let the ground be the bottom. Raise up the covers and put your nests in next to the ends. Use two brickbats to keep your nest material from working away. Put a shovelful of dust and coal ashes, equal parts, in each compartment next to the dividing partition, away from the nest end. Put water and feed as far away from the nesting material as you can easily reach toward the dust pile. This water should be changed every two or three days, according to weather. This gives you a light and cool place to set your hens and you can go along each day and raise up the covers and see if your hens and eggs are all right. Hens are never allowed out of the nest and run, as water and feed, both grain and green stuff, are kept in the little runs all the time. Also, a nice dust bath is handy to allow the hen herself to fight lice.

SOME LITTLE CHICK SUGGESTIONS.

Keep little chicks from getting chilled.

Never allow them to eat until 36 hours old, and until 48 hours only a little water and a small amount of hard-boiled egg.

Keep them as quiet as possible for the first three days, never allowing them more than a foot from the hover until after that time. A small plank partition may be put in the brooder in order to confine them close to the hover.

Little chicks may be taught to drink by sprinkling chick feed on top of the water, as they will peck it off and learn to drink.

An older chick, after it has had the run of the brooder, will be handy in teaching the chicks to drink as well as eat. To teach them to eat, sprinkle a small amount of feed on a white paper and tap the paper lightly with your fingers, placing the little chicks near the feed. They will soon learn to come when they hear this rattling noise on the paper.

When chicks are 8 to 10 weeks old, they should be taught to perch. Place the perches in the brood coop low at first and gradually raising them until they are the right height.

If kept two or three nights in a certain brood coop, they will learn to go to that coop at night, or the majority of them will, even when they have a larger range with the other chicks and the opportunity to go to another brooder.

Watch small chicks constantly for head lice; a large gray louse that is found on the head and neck, and is very fatal. A little grease or cream applied on the head and neck will take care of this trouble, but the chicks must be carefully preserved from getting wet while they have the grease on them.

FIRELESS OR LAMPLESS BROODERS.

Built From an Old Goods Box.

In recent years this system of brooding has become very popular, especially for indoor use, or where you have an old shed or in a climate where the early spring months are not subject to severe cold and rapid changes. They are not advisable for mid-winter brooding unless you have houses that will not get below freezing, and even with such houses in cold weather your chicks will show poor growth. For warm weather they can be recommended. The theory of a lampless brooder is that the chickens make enough heat to keep themselves warm, if it is only kept from escaping. This is not a correct theory for very small chicks, as they will become chilled, even when piled up in considerable bunches, if the weather is cold. You do not want chicks either chilled or piled up in bunches. The brooder, to hold 30 to 40 chicks, should consist of an outer covering 2x3 feet and the box 8 to 10 inches deep. The top should be in two sections, as per our illustration. The upper top should be hinged on to the rear. The strips of cloth shown in the illustration should be 2 inches wide and one-half of them 2 inches long and the other half 4 inches, and put in double rows around the four outer edges. They should be made of flannel or some soft material. Tack them in strips an inch apart, completely covering entire top. On warm nights it will be necessary to lift the upper covering some, especially as the chicks grow in age.

The brooder should be as simple as possible. Line it inside with roofing paper or double the walls. The cushion or hover part should be tacked on a separate cover three inches smaller each way than the top lid and thus will give better ventilation and allow the chicks to come outside if it is too warm under the hover. This cushion-holding frame need not be solid, but may be a frame of 1x2 stuff, covered with flannel or burlap. This covering on the frame may be allowed to sag in the center three or four inches. A cross-piece should be placed in the center of the frame, the long way. Tack the burlap to this. This sagged covering is nice for the chicks to brace their backs against. In using frame and cloth covering the hanging strips are tacked to the cloth with needle and thread. Place in the sagging burlap a bottleful of hot water at night and this will be found a help in keeping the chicks warm. Hot sand in a bag can be used in place of water, as it will hold heat longer.

The bottles of water or sand may be heated on a cook stove and then laid in the brooder. An extra cloth on top of the bottles helps retain the heat. The frame that holds the cloth or hover part should be hinged to the box about half way to the top, or it may rest on cleats nailed to the sides. By having the cleats nailed down low so that the hover when on the cleats fits the smallest chick, by putting in blocks of wood on the cleats, the hover can be raised as the chicks increase in size. An adjustable hover to fit chicks of any size may be hung onto the top lid by nailing an old broomstick just a little longer than your box is deep perpendicularly to the crosspiece of the hover frame. Bore a hole in the top just a little larger than the broomstick, passing the broomstick out through this hole. With a small gimlet make holes through the broomstick about 2 inches apart. Put a 10-penny nail through one of the holes, the one nearest the top for the least baby chicks, and the nail resting on the top holds the hover off the floor. By moving the nail you can regulate the height of the hover so that the inside of it just touches the chick's back.

These brooders may be used outdoors in a brood coop in moderate weather.

A heavy cloth over the entire brooder helps retain heat and keep out cold.

HOW TO TELL MALE AND FEMALE SOON AFTER HATCHING

In the single comb breeds the little chicks that are males have larger combs than the female chicks. Also some spurs can be seen on the little cockerels when only a few days old. The cockerels in the larger breeds are nearly always some larger. As the chicks grow the cockerels are inclined to be straight in body and do not feather out so evenly or quickly. The pullets put out small tails and wing feathers earlier than the cockerels. Cockerels sprout wattles, while pullets do not.

HOW YOUNG CHICKS LOOK.

In Black Minorcas and the majority of black fowls there is some white on them. White Wyandottes are inclined to have grayish colored chicks among them. In White Plymouth Rocks the chicks, frequently the best ones, will have a grayish appearance. Barred Rocks are dark brown with gray spots on back of head and some are nearly solid black. The Buff chicks in the Buff breeds are not rich buff, but are frequently nearly pure white. Brown Leghorns are marked with dark brown on a lighter background. Houdans are black on the upper part and a yellowish white underneath. Light Brahmas are yellowish white. Many black varieties have whitish throats and undercolored bodies. R. I. Reds are rich buff and sometimes approach red, but many of the good colored chicks will be nearly white.

Feeds and Their Preparation

"FEED FOR A FEW CENTS A BUSHEL."

Cheap methods of producing feed have been widely advertised and the process sold for a high price. There is no patent on the idea and you can utilize it.

The feed produced in this manner is especially valuable for winter feed, but will prove a good tonic for feeding yarded fowls even when other green food is convenient. You can feed freely of it. Sometimes the oats spoil and do not sprout properly. These spoiled oats must be fed sparingly, if fed at all.

The feed is known as "sprouted oats." The method can be practiced in warm weather in the poultry house or barn, care being taken to have the oats that you are sprouting kept away from the fowls until ready to feed. But in cold weather it is necessary to have a cellar for the process.

Here are some methods:

Take a box that will hold about a bushel and put the oats about 3 to 5 inches deep in it. Wet them good with warm water in the morning, stirring up carefully. Three or four small holes should be bored in the bottom of the box to allow surplus water to escape. The oats should be at least 2 inches deep and not over 6. It is necessary to have a warm place to sprout them. They should be stirred up and wet down every day. It will take two weeks to sprout them properly by this method. When sprouted they are a mass of green sprouts and oats.

Another way. Have a rack in the cellar, or wherever you wish to do the sprouting, with shelves or drawers, the drawers to be about 6 inches deep and the bottom of them of fine screen wire or burlap that will not let the oats run out. By making seven of these shelves or drawers and filling them one-third full of oats, and having them one above the other, you can throw the warm water on the top shelf and sprout feed for each day of the week, starting new shelves at the top. If the oats are not too deep and plenty of water is used, care being taken not to have the water too hot, it will not be necessary to turn the oats. It will be necessary in this method to soak enough oats over night to fill a compartment each morning, as pouring water through them will not get enough water in them.

A good method. Take 1 bushel of common oats and soak them over night. Make a frame of 1x6-inch lumber, 3 feet wide by 8 feet long. Place the frame on smooth, hard ground and spread the oats evenly inside it. Cover the oats with 1 inch of loose ground and water every day. When the sprouts show through, it is ready to feed. Then with a garden hoe, work under the roots, pull them up straight and you will have the finest green food you ever saw—food that you cannot beat with anything else as cheap or as simple to obtain. Every hoe full has a big lump of dirt on it; also some animal food. Examine some of this earth and you will find it full of worms and bugs. This gives the birds exercise to scratch them out.

Another way. Sow the oats in deep rows in the ground, soaking them over night before planting.

These last two methods can be practiced only during warm weather.

THE SCIENCE AND PRACTICE OF FEEDING.

The art of feeding is a scientific one, but should also be treated in a practical way. No man should give a hen anything to eat without having a reason for so doing.

What is he going to raise poultry for? Let us suppose it is primarily for eggs, and take up that standpoint. What is the egg composed of? The chemical composition of the dry substance of the inside of the egg is:

	Protein.	Fat.
White (albumen, white of the egg)	88.92	.53
Yolk	20.62	64.43

This composition would naturally lead us to think that in order to make eggs, fatty as well as nitrogenous food must be given.

A hen requires so much food per day as a maintenance ration. When this is supplied, and not until them, will she lay. All over this required amount is put into the egg until its limit is reached, then the surplus goes into fatty tissue. In order to lay, a hen must have some fat, the egg requires it, but not too much. An oversupply causes liver troubles, broken down legs, etc. Therefore, the problem is, feed enough, but not too much. But what is enough and what is too much?

Prof. W. P. Wheeler, of the New York Experiment Station, Geneva, N. Y., has found that 500 pounds live weight of hens in full laying, each hen weighing from 3 to 5 pounds, would require per day: Dry matter, 27.5 pounds; ash, 1.5 pounds; protein, 5.0 pounds; carbohydrates, 18.75 pounds; fat, 1.75 pounds. This has a nutritive ratio of 1:4.6. The nutritive ratio means the ratio between the carbohydrates, plus 2¼ times the fat, and the protein. The protein makes the fibres and muscular tissue, and the carbohydrates and fat produce the fatty tissue, energy and heat.

We know how much food constituents it takes to keep a hen in laying condition and what the analysis is of the egg after being laid. So what grains, etc., shall we feed? Fowls being primarily grain eaters, we discuss the principal ones.

Wheat.—The best one grain is palatable, small size, not hard and is rich in mineral matter as well as protein. It is an easy grain to obtain and should be used extensively in the feed mixture. Screenings are good if clean.

Corn.—Is the king of grains, because it is so cheap and economical, except when fed as a single grain. Alone, it is too fattening, does not contain enough protein, but does have too much mineral matter. Hens will pick it out in preference to all grains, partly on account of its color; corn gives a yellow tinge to the yolk of the egg and also to the plumage of the fowl.

Oats.—If heavy, are desirable. The grain is a growthy feed and the oat oil is supposed to contain avanin, a nerve tonic.

Barley.—Barley is good for fattening, and if handy and reasonable in price, is equal to wheat. Hens like it.

Buckwheat.—That is used for fattening, and whitens the yolk of the egg and skin of the fowl. The middlings are richer than wheat middlings.

Rye.—Do not feed. Hens rarely like it.

When forcing hens to lay, their work of digestion must be reduced, hence we feed ground food made mostly of by-products. Ground food is used to narrow the ration because no combination of whole grains is narrow enough for a laying hen.

Bran.—This is best for bulk and can be fed liberally. There is a big difference in quality and the test is to chew it. If gummy, it contains gluten, which is the valuable part of bran.

Middlings.—This is sometimes made of ground bran, but flour middlings are best. When mixed with water, it should be sticky. A low grade of flour is better than most middlings. But it is a valuable food in a dry mash.

Linseed Meal.—This is a laxative and should be fed sparingly, never more than one-tenth.

Cotton Seed Meal.—Never feed. It is constipating.

Meat Foods.—Give a liberal amount if the hens are confined, for there is no substitute for it to be found in grains. The best is skim-milk. Curdle the milk and feed the pot cheese.

Green Cut Bone.—This ranks next, but is sometimes unhandy to feed. Never feed more than ½ ounce per hen per day.

Beef Scrap.—Much illness is caused by feeding this, especially if the quality is poor. Get a guaranteed analysis and test by smelling the aroma given off when hot water is poured on it. But it is a valuable feed and can be fed as 10 per cent of the feed.

Green Feed.—If confined, feed the hens green feed, as it aids in digestion. Mangel beets are the best. Rape and alfalfa are good. Alfalfa meal may be fed in winter.

Now that we have looked at some of the important feeds, how shall we feed them? First, let me say: Keep as close to nature as possible. A hen eats when she wants to, from daylight until dark, and all she wants of it. So we must allow this, but possibly control what he shall eat. If allowed free range the feed bill will be greatly lessened and the fowls probably much healthier, but everybody cannot allow this.

Some of the points in a good ration for fowls are:

1. It should be composed of foods every one of which the fowls like. Do not try to force them to eat undesirable food.

2. It should contain a sufficient quantity of digestible nutrients to supply the needs of rapid growth and large production.

3. It should have enough bulk to enable the digestive secretions to act quickly upon it.

4. It should not contain an excess of indigestible fibre, which must be thrown off by the system, thus causing a waste of energy.

5. A certain portion of the ration should be of whole grain in order to provide muscular activity of the digestive organs. This is made necessary in grinding the grain. Under certain conditions a quantity of the ration should be soft, ground food. This is for the purpose of providing quickly available nutrients to supply the immediate demands of rapid growth or heavy continuous egg yield.

6. It must provide a good variety of foods, in which are included grain, green food, meat and mineral matter in proper proportions.

7. The foods in the ration must not have an injurious effect upon the color or flavor of the product.

8. The age of the fowl, the breed and the kind of product it is desired to produce must be considered, as to whether the food is intended to grow muscle and bone, or to produce eggs, or to fatten.

9. The ration must provide the two classes of food nutrients; the protein and carbohydrates, in such proportion that they will supply the daily needs of the fowl's system; it must also provide sufficient digestible protein to repair the waste tissue with new growth and to produce eggs and provide the proper amount of digestible carbohydrate food to furnish heat, energy and lay by a little surplus flesh in the form of fat.

10. The ration must consist of foods which furnish the nutrients at the lowest possible cost.

11. It is not how much a fowl eats, but how much it can digest, that determines the value of a food. Various classes of animals differ in their power to digest the same kind of food. Foods also vary in their digestibility when used by the same animal. Since the proportion of each poultry food which fowls can ordinarily digest has not yet been determined, we will accept the standards of digestibility which are used in compounding rations for other animals.

It has been our experience that a hen should be made to exercise if she will not do it naturally. To do this a grain feed should be put in a straw litter. Early in the morning a few handfuls may keep the hens busy all day. At night, especially in the winter, they should go to bed with full crops. If we can keep the ration narrow enough, all the feed a hen wants will not overfatten her, if she exercises.

As has been stated before, all grain is too fattening, so ground feed must be given. Wet mash takes too much time for the results obtained, so think the dry mash is best. Make a good hopper that is non-wastful, and keep before the fowls all day. If you care to, close them at night.

Grit should be before them at all times, also oyster shell. Grit is the hen's teeth; oyster shell makes the egg shell; so both are necessary.

Green food should be given, and water is indispensable.

The following ration has been found to be a successful one, where eggs are in demand:

The proportions of grain mixed together for the litter is 10-

10-5: Wheat, 10 pounds; corn (cracked preferred), 10 pounds; oats, 5 pounds.

The proportions of grain mixed together for the litter is 10-10-5: Middlings, 6 pounds; corn meal, 6 pounds; bran, 3 pounds; oil meal, 1 pound; alfalfa meal, 1 pound; beef scrap, 5 pounds.

This nutritive ratio is 1:3.9, a little narrower than Professor Wheeler says is necessary. In feeding, the grain and ground food in the above proportions should all be eaten up at the same time. Hens will naturally eat more grain than meal, so the grain may have to be held back in order to make them eat the meal. Generally the grain will run out just a little ahead of the meal, so the ratio runs about right most of the time.

Now, do not think we mean that all hens should be fed grain in the proportion of 1:4.6, because in the hen business too much scientific work is a failure. But the above discussion of feeds is a good guide. Any poultryman should use his own judgment, and suit his actions to his conditions and environment, etc.

With some fowls the above ration is a grand success. With others, we find all the mentioned feeds cannot be obtained, so the next best thing is done.

We have talked to feeders who know nothing of the science of feeding, but just use common sense. They claim a hen is like a man—she wants a variety and she wants a plenty. But the real and main point is to know the hen and give her what will bring the best results.—A. G. Philips.

A FAMOUS CALIFORNIA EGG FEED.

This ration, fed with green feed of potatoes, cabbage, etc., gave the highest results in egg yield at the California Experiment Station. It is fed in hoppers as a dry mash.

Two parts middlings, 2 parts bran, 1 part corn meal, 1 part shorts, 1 part bolted barley, 1 part meat meal, ½ part bone meal

A KANSAS CHICK RATION.

Grain mixture. Two parts cracked corn, 2 parts cracked wheat, 2 parts Kaffir corn, 1 part millet.

Mash. Two parts corn meal, 2 parts shorts, 2 parts bran, 2 parts beef crap, ½ part alfalfa meal.

Give them all the grain and mash which they desire. As they grow older, change the mixed grain to a mixture of whole grain rather than cracked grain. This will furnish variety and practically all the necessary things which grain will furnish a growing chick In feeding the mash, which feed in hoppers, be very careful that they do not overeat when quite young, as some chicks are apt to become overzealous in their desire for this good tasting mixture so that they become affected with diarrhoea and consequently are held back somewhat in their growth, but with judgment this combination of feeds is safe.

This ration has been very successfully used by the Kansas station, and the loss of chicks was practically nothing. It produced very quick-growing and thrifty chicks.

HOW MUCH A HEN EATS A YEAR.

The average amounts of the materials eaten by a hen during a year is about as follows: Grain and meal mixture, 90.0 pounds; oyster shell, 4.0 pounds; dry cracked bone, 2.4 pounds; grit, 2.0 pounds; charcoal, 2.4 pounds; clover, 10.0 pounds.

WEIGHT OF POULTRY FEED.

Frequently it is desired to weigh up feed rather than measure it. Below are the average weights of the most commonly used poultry feeds: One quart middlings, 1 pound; 1 quart shorts, 1 pound; 1 quart bran, ¾ pound; 1 quart alfalfa meal, ¾ pound; 1 quart rolled barley, 1½ pounds; 1 quart wheat, 2 pounds; 1 quart corn, 2 pounds; 1 quart beef scraps, 1¾ pounds; 1 quart beef or blood meal, 1¾ pounds; 1 quart oyster shell (crushed), 3 pounds; 1 quart limestone grit, 3 pounds; 1 quart millet seed, 1¾ pounds; 1 quart unshelled oats, 1 pound; 1 quart charcoal (crushed), ¾ pound, and 1 quart Kaffir corn, 1¾ pounds.

MAINE STATION CHICK FOOD.

Infertile eggs are boiled for half an hour and then ground in an ordinary meat chopper, shells included, and mixed with about six times in their bulk of rolled oats by running both together. This mixture is the feed for two or three days, until the chicks have learned to eat. It is fed with chick grit, on the brooder floor, on short clover or chaff.

About the third day the chicks are fed, in low tin pans, a mixture of hard, fine broken grain, as soon as they can see to eat it in the morning. The mixture now used has the following composition:

	Parts by Weight.
Cracked wheat	15
Pinhead oats (granulated oat meal)	10
Fine screened cracked corn	15
Fine cracked peas	3
Broken rice	2
Chick grit	5
Fine charcoal (chick size)	2

When they are about 3 weeks old the rolled oats and egg mixture is gradually displaced by a mixture having the following composition:

	Parts by Weight.
Wheat bran (clean)	2
Corn meal	4
Middlings, or "red dog" flour	2
Linseed meal	1
Screened beef scrap	2

This mixture is moistened with water just enough so that it is not sticky, but will crumble when a handful is squeezed and then released. The birds are developed far enough by this time so that the tin plates are discarded for light troughs with low sides. Young chicks like the moist mash better than that not moistened and will eat more of it in a short time. There is no danger from the free use of the properly made mash twice a day, and since it is already ground the young birds can eat and digest more of it than when the feed is all coarse. This is a very important fact and should be taken advantage of at the time when the young chicks are most susceptible to rapid growth, but the development must be moderate during the first few weeks. The digestive organs must be kept in normal condition by the partial use of hard feed, and the gizzard must not be deprived of its legitimate work and allowed to become weak by disuse.

By this time the chicks are 5 or 6 weeks old and the small broken grains are discontinued and the two litter feeds are wholly of screened cracked corn and whole wheat.

MAINE STATION FEED FOR BREEDING STOCK.

Early in the morning for each 100 hens 4 quarts of whole corn is scattered on the litter, which is 6 to 8 inches deep on the floor. This is not mixed into the litter, for the straw is dry and light, and enough of the grain is hidden so the birds commence scratching for it almost immediately. At 10 o'clock they are fed in the same way 2 quarts of wheat and 2 quarts of oats. This is all the regular feeding that is done.

Besides the dry whole grain a dry mash is kept always before the birds, composed of:

	Parts by Weight.
Wheat bran	2
Corn meal	1
Middlings	1
Gluten meal, or brewers' grains	1
Linseed meal	1
Beef scrap	1

AN ALL AROUND FEED.

A good feed for breeding stock is made as follows: Cracked corn, 30 pounds; whole wheat, 30 pounds; oats, 10 pounds; clover or alfalfa (fine cut), 10 pounds; beef scraps or meal, 10 pounds grit, 4 pounds; oyster shells, 4 pounds; charcoal, 2 pounds.

Feed as a dry mash in feed hoppers. Green feed should be given in addition. If the flock seems to get too lazy or fat, close the hopper and feed grain in litter for a week.

A QUICK GROWING RATION.

After the chicks are 6 weeks old and have been fed one of the rations recommended in another place, the following ration may be fed them in hoppers: Five parts wheat, 4 parts cracked corn, 3 parts of mash composed of 45 pounds corn meal, 10 pounds Kaffir corn, 10 pounds ground oats, 5 per cent beef scrap with 1 pound grit and granulated bone. Keep this before them all the time and allow them range, and your pullets will be laying in five months from hatching. They will also develop thrift and size.

A VERY STIMULATING FEED.

An excellent ration is made by mixing 100 pounds corn chop, 100 pounds of bran, 100 pounds of shorts, 100 pounds alfalfa meal, 50 pounds oil meal, 50 pounds meat meal and mix in this a pint of salt and 2 pounds of cayenne pepper. Keep this in your feed hopper at all times. About a half hour before sunset feed your chickens all the whole corn they will eat. Be sure they eat it all up clean.

This feed is very high in egg-producing material and will prove a fine tonic to start a flock laying. If it causes bowel trouble, take the mash away for a few days, feeding corn and wheat only.

FATTENING FEEDS.

A much heralded fattening feed is made of equal parts clover or alfalfa meal, corn meal or wheat middlings and bran. The best proportion, perhaps, is one-third each alfalfa meal, corn meal and wheat bran. The mixture is scalded well with hot water and stirred up thoroughly and let stand until cool for feeding. It is claimed that an addition of 10 per cent tallow will increase the value of this as a fattening food.

The Maine station fattening recipe consists of 100 pounds corn meal, 100 pounds wheat middlings, 40 pounds meat meal. This is fed with a mash condition, just thick enough to run off a spoon. If wet with skimmed milk it will be much better than when wet up with water.

Here's another fattening mixture for young stock especially: Equal parts corn meal, ground oats and wheat middlings, with a little linseed meal. Feed all the whole corn they will eat once a day.

Experiments have shown that the greatest gain in fattening is made by young stock from 3 to 4 months old.

Fatten young cockerels and sell at this age.

Old stock may be fattened for market on either of the above feeds.

CRATE AND PEN FATTENING.

In the crate method a few fowls are confined in crates and fed from a trough. A crate 6 feet long, 18 inches high and 18 or 20 inches wide is suitable and is large enough for a dozen birds. Sometimes such a crate is divided into two or three compartments, 4 to 6 birds being placed in each compartment. But little room for the birds to move about in is desirable, for the less exercise a bird obtains the more rapidly does it fatten. The top, back and ends of the crate should be solid if they are to be placed outdoors, but if they are to be in a building, they may be built of laths or slats. These slats should be 2 inches apart in front, so as to permit the birds to eat from the troughs which are hung just outside the coop. The slats of the bottom of the coop should be about 1 inch apart to permit the droppings to fall through. In indoor feeding the crates should be placed in a dark room, and just before feeding enough light should be admitted to allow the birds to see to eat. They are usually fed three times a day, and are permitted to eat for half an hour at a time, when the room is again darkened and the uneaten feed removed.

It is claimed that crate or coop fattened poultry is more tender than that fattened in a small house or allowed to range, but experiments have demonstrated that chickens fattened in a small house will make greater gain than those fattened in crates.

Twenty-five fowls may be fattened in a room 10x12 feet, and will make greater gain in two weeks than those fed in crates.

Grit and plenty of water are kept before fattening fowls all the time.

In crate or pen fattening, many leading poultry buyers have lately adopted the plan of feeding from 7 to 10 days only, as they claim it does not pay to feed a longer period.

Preparation for Exhibition

HOW TO CATCH AND CARRY POULTRY.

Two methods are practiced by fanciers in taking poultry out of exhibition coops, and both are good. The writer prefers the first one.

Reach in with your right hand, placing it on top of back and wings of the bird; press down carefully, yet firmly and gently, until the bird begins to give from the pressure, then take hold of the legs with the left hand, grasping the wings with the right hand. The thumb and forefinger should go around the right wing of the bird, while the remaining fingers grasp the left wing if an attempt is made to spread them. Remove the bird head foremost from the coop.

Some, however, practice the opposite and remove the bird tail first. Never attempt to bring the bird out sideways.

In returning the bird to the coop, always return it tail first and it will not make an attempt to get loose before you get it well inside the coop. By handling a bird this way, holding it firmly, there is the least liability of damage to the plumage.

Another method is to catch the bird by the feet first with the left hand, allowing the bird to rest on the forearm, and the right hand can be used if necessary to hold the wings.

In handling wild birds, it is necessary to work slowly and with caution, but with a few handlings the average bird submits and can be handled quickly and easily.

In carrying any kind of poultry carry them by the feet with the head up, allowing the body to rest on the forearm and the head under the arm and next to your side. Never carry poultry feet up and head hanging down. Also, if you carry a bird by the feet with its head out in front of you, it will be likely to become frightened and attempt to get loose, but with the head under the arm it will submit quietly.

FAKING OR FITTING?

In the realm of poultry exhibitions and in the opinion of fanciers there is a dispute about where legitimate fitting of a bird for exhibition ends and where disreputable faking begins.

The Standard describes faking by naming quite a number of things, and then closes with the following definition that may be taken as general: "Any self-evident attempt on the part of an exhibitor to deceive the judge and thus obtain an unfair advantage in competition."

It sounds like this would be general and sweeping enough to cover and settle all disputes, but still it is a fact known among all well-posted poultrymen that many of the leading exhibitors do things that they would not want the judge or their competitors to know. They will defend some of these by word of mouth and pen. It is not intended to say who is right or wrong in these disputed questions of just where fitting quits and faking begins, but to give the facts and let the reader judge for himself.

The question separates itself into three distinct divisions and will be treated in this manner.

First. The unobjectionable or admittedly proper fitting.

Second. The debatable or questionable things: call them "fitting" or "faking," as you prefer.

Third. The straight-out faking.

PROPER FITTING.

Some of the methods of fitting that are pretty generally recognized as all right and that you may do with a good conscience will be described.

Cleaning Legs and Toes.—All birds that you are going to enter in the show room should have legs and toes carefully washed and cleaned a few days before taking them to the show. Use strong soapsuds, rubbing the feet and legs with a stiff-bristled toothbrush. Get all dirt out from under the scales, even if you have to take a tooth-pick and clean the scales one at a time. If there is scaly or knotty formation on the feet, you may have to first grease them carefully, removing the knots and then washing next day. After carefully washing and cleaning thoroughly, rub with a little sweet oil, in which has been dropped one or two drops of carbolic acid. After washing you will have to keep the birds out of the dirt by keeping in a clean place, or your work will all have to be done over again.

Polishing Beak and Toe Nails.—In some breeds the appearance of the birds may be very much improved by carefully polishing beak, as well as toe nails. This is true in Barred Rocks and breeds that are inclined to show dark spots on beak. Frequently the objectionable discoloration can be polished off without injury to the bird. Care must be taken not to polish to the quick, so as to injure the bird and make the beak tender.

Oiling Comb, Etc.—The comb, wattles and face can be heightened in color by using a little glycerine and sweet oil. Alcohol may also be used. This should be used only a short time before judging, as it brightens up the comb, wattles, etc., very much. After rubbing either on, take a cloth and carefully clean off all the fluid, leaving comb, etc., bright.

Preventing Comb From Falling Over.—Frequently birds taken from an outdoor life into a warm show room, especially where they have been well cared for, the combs of the high single comb breeds will be inclined to slightly fall over, and thus materially injure the appearance of the bird and even hurt its value in the eyes of the judge. This may be remedied by washing the comb a few times in cold alum water.

A method of making a wire frame is sometimes practiced to support a comb before the show. Two pieces of pasteboard may also be fastened on each side of the comb and support it. It is usually difficult to get a bird to allow its comb to be supported in this manner.

Training Birds to Pose.—It is important that your birds be gentle and easily handled when you take them to the show. About two or three weeks before the show, they should be kept in exhibition coops, so they will not be excited at being confined. They should be handled a few minutes every day. Begin handling them at night, as you will find them much more docile at night. Take the bird out some night when you have only a moderately good light and stand on a barrel or box, talking to it and handling it all the time. Raise its head up gently by touching point of the beak, then press it back down, move the bird slightly a few times, all the time keeping your other hand moving on the back, wings and tail. If the bird is excited, handle it only a short time, trying again the next night.

After you get the bird so it will stand quietly in the night, try it in the daytime. Then begin teaching the bird to pose in the ideal station illustrated in the Standard. You can train the bird to raise or lower different parts of its body, so as to "even up," so far as you can, any deficiency that nature has made along this line. A few hours' work for three or four days will do wonders in teaching your bird to present a pretty picture to the judge.

Birds that are going to be exhibited should be trained with a judging stick so that they will not be afraid of one when the judge inserts it in the coop to shape them up. A common yard stick, pointer or a small stick about three feet long will do. After the birds have learned to handle well, this stick should be taken and bird touched on the back and breast while standing in the coop. At first it will be afraid of it, but gradually will learn that when the stick is placed near it that it should stand up and assume typical position, carry its head in the air and pose.

The Straw Method of Washing Fowls.—Some fanciers clean their birds' plumage by what is known as the straw system. That is, four or five weeks before the bird is wanted for the show, he is confined in a house with a board floor, or large exhibition coop, with from 12 to 18 inches of straw all over it. This straw should be the very cleanest and absolutely free from dust. By keeping the birds in this straw and changing it every three or four days, according to the size of room and number of birds in it, so as to keep it fresh and clean, the plumage will of itself become clean and take on a very brilliant polish. This method is considered very tedious by the majority of fanciers and the most of them wash their birds with water.

Polishing Plumage.—Many fanciers wash even the dark and parti-colored birds, as they claim it gets the dirt out of the feathers and improves the appearance. The dark breeds, such as Brown Leghorns, Barred Rocks and Buff varieties, are very much improved in appearance of color by polishing the feathers just a few hours before they are exhibited—better still, just before the judge examines them. An old silk handkerchief, as soft as possible, is the best thing to polish them with. Take the handkerchief in the hand and rub the feathers downward towards the base of the tail. Care must be exercised not to break the web of the feathers.

How to Wash and Blue for the Show Room.—"The first thing necessary is to prepare a kitchen for the work and start a good brisk fire in a wood stove. Warm plenty of clean rainwater and set three good sized tubs close to the stove. One needs a good assistant, and no better can be found than a good, patient woman. First, have all the birds ready, so that no inconvenience will be caused by having to go to the hen house just when they are wanted.

"Tub No. 1 should be half full of water, as near blood heat as possible or a little warmer will not hurt. Put the bird gently into the water, holding it there either by the feet or by the sides of the body. If the bird has never before been washed it will not know what to make of it at first. Take it easy for awhile; hold the bird down in the water, partly immersed, and in a little while begin applying water with a good sized sponge. With this rub the feathers with the web as they lay; never rub against the lay of the feathers. The best way to hold a bird in the water is, when seated on a chair close to the tub, to face the birds towards you and wash away from you. After applying water with the sponge for awhile, then begin using the soap. Use best castile soap.

"Rub with the sponge and soap, turning the feathers over and over.

"Tub No. 2 should also be half full of clean, lukewarm water. Here the bird is put, as in tub No. 1, and washed thoroughly, so as to get out all the soap. After rubbing with sponge and using clean water freely, take a dipper and keep pouring the water out of the tub over the bird, letting it fall with a little force from about a foot above the bird. Do this all over the bird. If you do it right, one tub of this kind of work will be enough.

"Tub No. 3 should have in it some cold water, with just as much bluing as the good wife uses for bluing the white clothes. Into this the bird is put and rinsed with the cold bluing water. The water should be just cold enough to be chilly. The reason for this cold shower is for the same purpose that men take them after taking a plunge in a warm bath—it prevents catching cold.

"After taking the bird from tub No. 3, put the bird on a board placed on top of the tub and by means of the hands squeeze all the water out of its feathers you possibly can. Remove the bird then to the top of a box, or a chair, placed very close to a good brisk fire, and begin toweling with warm, dry towels, so as to absorb all the moisture out of the feathers that you can.

"Keep it before a brisk wood fire, but not so near as to curl the feathers or you will spoil them. With a strong palm fan let the assistant begin the drying, first fanning one side and then the other. This part of the work is gone on with until the bird is thoroughly dried. The fire needs to be brisk, the bird kept turned around and the fan going all the time, and it is surprising how soon the feathers will dry and open up so nice and fluffy. The fluff, the back and under the wings will be longest in drying. Holding a wing up with one hand, and fanning with the other, will soon make wonderful changes. If the work has been successful thus far, little difficulty will be experienced in getting the birds dry and putting on the finishing touches.

"It is a difficult job to get the feathers wet and soapy, and a difficult one to get the soap out again. But if, on drying, it has been found that the feathers are sticky and do not open nicely, which will not be the case if the work has been done right in the first place, put in order again another fresh lot of clean, warm rainwater, and rinse over again as before and proceed with the bleaching and drying again."

THE DEBATABLE OR QUESTIONABLE THINGS.

The things here given are many of them disputed questions among fanciers and they are not run for you to practice, but are given so that you may know and detect these if used by your competitor or used on birds sold to you.

Pulling Out Off-Colored Feathers.—Very frequently breeders of white breeds, especially where there are just a few feathers with dark specks on them on back, body and fluff where not disqualification, are removed and some even remove them where they are disqualifications. In White Wyandottes, White Rocks, White Leghorns, etc., it is considered good fitting to pick out the feathers on body, back and fluff that have gray specks in them. Of course, primaries and secondaries in wings cannot be removed without detection. This may not be according to the views of some, but if you will visit the yards of many of the leading fanciers you will see this work going on. In Barred Rocks removing black feathers or very badly barred feathers on back, body or fluff is frequently done by fanciers who are considered the very best. In fact, you will frequently hear a fancier state that he lost out by simply missing a black feather on his bird.

In the Buff varieties, R. I. Reds, etc., dead feathers, past matured, are frequently removed from back, wing bows, etc., as are also feathers that show shaftiness or some meliness.

Feathers are frequently removed from parti-colored varieties, such as Golden, Silver and Partridge Wyandottes, Polish, Hamburgs, etc., so as to allow the well-marked feathers to appear more distinctly. Feathers that are imperfect in coloring are taken out and the well-marked one left.

Badly colored feathers are frequently removed a few weeks before the exhibition just so other feathers will be grown out about half way by show time. This is done especially where the feathers are off-colored down next to the body of the bird. If many feathers are removed, the extra ho   rd usually attracts the attention of the judge and he will penalize accordingly.

Burning Off Defects.—Taking a cigar or match and burning off the white end of Barred Rock feathers, especially on back and breast, is sometimes practiced. It is a treatment that takes a good deal of time, especially if there are many feathers with this defect. It also requires considerable care, but it can be detected, as the burnt off feathers will not end so smoothly or velvety as the natural feathers. A miscroscope will frequently reveal this defect, although after treating a feather this way the burnt portion is rubbed carefully between the fingers to remove the charred part. White feathers or where there is just a little gray are frequently treated in this manner in black, buff or parti-colored birds. If the defect is very deep in the feather, of course, it cannot be treated without detection, and can be detected by the miscroscope showing the blunt burnt off feathers.

Treating Defective Tails.—Tails that are slightly wry and some even wry enough to be disqualified are sometimes treated by tying a weight to the tail so as to remedy the defect. Squirrel tails or those carried too high are also lowered by weighting them down and sometimes the feathers are even dampened with hot water and bent down to proper shape. The dampening and bending process is looked on with more disfavor than the other even.

Making a Spike.—In rose comb varieties, where there is no spike, and the absence of spike is a disqualification, an attempt to remedy the trouble is sometimes made by cutting around the base of the spike with a sharp knife and press the spike out until it extends beyond the base of the head. This is a very difficult operation and very hard to successfully perform. It is much harder to perform in old fowls than young ones and the younger they are treated for this trouble the more likely of success and the treatment not being detected. This is so questionable an act that it ought to be rated under faking, but some very good fanciers practice it and defend the process.

Use of Prepared Chalk, Corn Starch, Etc.—After birds have been washed in white varieties of prepared chalk, corn starch or talcum powder are sometimes liberally sprinkled in the plumage. Where this is done just before the bird is sent to the show the starch or other preparation will largely work out through the feathers and help increase the whiteness. This is pretty generally recognized as legitimate, also critcised by some as a doubtful method of fitting.

Bleaching by Means of Chemicals.—In white varieties quite a few fanciers practice what is known as bleaching, or "peroxiding." There are several methods of bleaching white birds, or taking the sap, as some call it, out of the feathers. Those who practice this method say it is no more changing the natural color than is using a little bluing to whiten white birds. On the other hand, those who oppose the method claim that it gives the feathers a dead white effect that is not natural, while bluing gives the natural effect. Bleaching can be detected often by the lack of luster in the plumage of the fowls. The white does not seem to have the same rich effect as when blued.

Wash your birds through two waters, as described in another place, but do not use any bluing. Dry the bird well. After well dried, take peroxide of hydrogen and wash any sappy or brassy feathers you find on back, wings, neck, hackle and tail. Take the pure peroxide, using a soft tooth brush to rub the peroxide into the feathers that show "off" color. Be careful to get the feathers well wet. After making sure every brassy or sappy feather has been touched with peroxide, place the bird in a coop with plenty of straw in the bottom to dry and bleach. The coop should be in a moderately warm room and the bird not too near the fire. After the bird is well dried, take prepared chalk and polish carefully the outside feathers to give them a silky gloss. A bird is usually "peroxided" a short time before taken to the show room, and kept out of the sunshine until the show.

If all the sap is not taken out the peroxide is applied again after waiting a day.

A small amount of ammonia may be added with good effect to the process. Not over 1 per cent should be used and care taken not to get the mixture into the eyes or too much on the flesh. If the flesh is irritated and reddened, the bird should be washed in warm water, even if the "peroxiding" has to be repeated in a few days.

Some wash the entire bird, feathers, feet, comb, etc., in pure peroxide. It is rather irritating to tender parts like the eyes.

STRAIGHT-OUT FAKING.

The Standard of Perfection names as faking, besides the general description given, the following:

Removing or attempting to remove, foreign color in face or ear lobes when it is a disqualification.

Removing one or more side sprigs, or trimming a comb in any manner, except the dubbing of Games.

Artificial coloring of any feather.

Splicing feathers.

Injury to plumage of any fowl entered by another exhibitor.

Plugging up holes on legs of smooth-legged varieties when feathers or stubs disqualify.

Staining of legs.

Under this head may be classed, also, the following not named in the Standard:

Entering fall hatched hens or cocks at shows the fall following as hatched in that year.

Entering birds over a year old as less than a year old.

Doctoring Ear Lobes.—A common test used among poultry judges to tell if white in ear lobes is permanent is to rub the ear lobes and then turn the bird with the head down for half a minute or more. If the rush of blood to the head removes the white in ear lobes the bird is counted all right and not subject to disqualification for the defect. Some fanciers burn out white in ear lobes, where it is objectionable, with a weak solution of lunar caustic. This is generally very easily detected, as it leaves a scar showing where the caustic has been used.

Frequently an attempt is made to remove red from white lobes in the Mediterraneans, etc., but this is very difficult to perform.

The lobes are sometimes burned with caustic, but this generally makes a bad matter worse.

Birds out of condition are more likely to show off color in lobes than those well fed and cared for. A stimulating tonic, even a few drops of whisky or some pepper for a few days, will help the color of red lobes that are too pale. Confinement in warm show rooms will often cause birds to lose color of lobes. This can be improved by rubbing red lobes in a little alcohol a short time before the judging.

Side Sprigs.—One of the greatest temptations to fake a new fancier has is when he has a "cracker-jack" bird in all other respects except a small side sprig. He feels that he can take a razor and clip off the offending sprig very easily, and if done a few weeks before the show the injury will heal up. Or he has a bird with too many points in comb and he clips out one.

This species of faking is more objectionable, perhaps, when a side sprig is removed, as this is a disqualification, than where the point is cut out of a comb, as this is simply increasing the bird's score.

Trimming a comb is very easily detected, as the part where the cutting is done shows smoother and does not have the fine, granulated effect that the other parts have. The earlier the comb is doctored, the less likelihood there is of detecting it. Often cuting a poin out of a comb, or removing a side sprig, cannot be detected if the doctoring was done in the first two or three months of the bird's life. Various methods have been tried to produce the natural granulated effect, but none is very successful. Some of them may be mentioned as burning with lunar caustic after the cut, greasing with salty grease, etc. Judges are very strict on the question of trimmed combs, and disqualify promptly for evidence of it. Usually the only evidence needed is the doctored comb.

Artificial Coloring.—Artificial coloring of the feathers is a very rare fake, as it can usually be detected very easily. This is resorted to in wing and tail feathers more often than anywhere else. It is not often attempted at all, as the slightest difference in shade of coloring is easily detected by the naked eye and thus puts the bird in the disqualified class and its owner in disgrace.

A famous case of coloring plumage is the one where R. I. Reds were colored with Diamond dyes and exhibited at the Jamestown Exposition. The faking was very plain and the birds were thrown out. Their owner was also expelled from the R. I. Red Club for this faking. Another noted case of more recent occurrence, the primary feathers in the wing were painted black in a Red with India ink This was detected by an exhibitor rubbing the ink off. The exhibitor who showed the doctored bird was expelled from the American Poultry Association and R. I. Red Club.

Splicing Feathers.—This is attempted very rarely, and then only in main tail feathers and in wing primaries and secondaries. The object, of course, is to put in a good colored feather out of some other bird, removing the bad feather from the bird doctored. This species of faking can be detected very easily, the only thing being that a clever piece of splicing might not be noticed, but a close examination of the wing and main tail feathers readily reveals it. The spliced feathers are usually fastened in with fine wire or glue.

Injury to Plumage of Another's Birds.—Of all methods of faking this is the most disreputable. This charge is occasionally heard around the show room: "Somebody broke two feathers out of my best bird's wings." Or you hear the wail: "Somebody has pulled out one of the sickle feathers from my best male."

This occasionally may be the work of a rival exhibitor, but frequently these feathers were broken while being shipped and no one has been guilty of such disreputable conduct as injuring another's birds.

Stubs, Etc., on Legs.—The Standard describes as faking "plugging up the holes of smooth-legged varieties where feathers or stubs disqualify." The inference is that the owner of the bird has removed the stubs and sought to fill up the holes with beeswax or soap, they being the most commonly used, and about the color of the average fowl's legs. Stubs or feathers can sometimes be removed from the leg and the fake not detected, but generally, if the judge has a sharp eye, he will see the holes left. Some breeders remove the stubs just a few days before bringing the birds to the show and then with a sharp knife trim around the scales on the legs so that the holes are not quite so conspicuous. Faking by removing the stubs or feathers on the legs is too commonly practiced and frankness compels the acknowledgment that it is not looked on with as much disfavor as it perhaps should be. This is largely because there is a big school of fanciers who believe that feathers on legs should not be disqualification, but should simply be entitled to a cut. These claim that there is no harm to remove these little disqualifications and their conscience apparently backs them up in the removal.

Many a fancier has, as he thought, removed all stubs only to have his clever work discovered in the show room and his bird put out of the winning. No doubt he was correct in thinking he had all stubs out when he started with the bird to the show, but feathers on the feet seem to grow remarkably fast and four days are enough to develop a nice growing stub. It is claimed by some that small down will grow on an extra vigorous bird in two days.

Staining Legs.—This is seldom resorted to, even by the most reckless fancier. It is seldom necessary; in fact, the average bird, with even worse than the average pair of legs, can have the color very much brought to its legs by washing them carefully in soft water and soap, then greasing them with sweet oil or vaseline. Applying these remedies is not counted as staining. A case of staining, unless very carefully performed, can generally be detected by careful examination, as the stain, unsually butter color or analine dye, is not worked in under the scales thoroughly. Also, the legs are usually stained too dark.

Misrepresenting Age of Birds.—Frequently an exhibitor enters a bird hatched in January or February at a show the next fall or early winter as a hen. In Buff varieties and other varieties that are liable to fade with age, this, of course, is reprehensible, as the young bird holds its color better. This is very hard to detect. However, the information given in another chapter about telling the age of fowls will prove helpful. Occasionally this order is reversed and an old bird entered as a young one, especially December hatched cocks being entered as cockerels.

This deception is hard to detect positively, although fanciers frequently draw pretty accurate conclusions, even despite the owner's protest as to age of his bird.

CURING SOME BAD HABITS.

Breaking Hens From Setting.—One of the most persistent habits to break, especially in the late spring, is that of setting hens when you do not want them to set. The most important thing in breaking a setting hen is to get started early. After a hen has set a few days she is much harder to break than if taken off the nest the first day. There is also a great difference in individual hens of the same breed. The Mediterranean classes, including the Leghorns, Minorcas, Hamburgs, etc., are non-setters. Occasionally one will set and prove a good mother. However, they are considered unreliable, as they will set a few days and then leave the nest for no apparent cause. The Asiatic classes, Brahmas, Cochins and Langshans, are more persistent than the American classes—Wyandottes, R. I. Reds, Plymouth Rocks, etc. Cochins are perhaps the most persistent of the breeds.

Setting hens lose weight continually while setting and also while you are trying to break them. If you wish to cull out setting hens, as soon as they have quit laying, the first day they go to setting catch them up and take them to market.

What will succeed in breaking one hen will often fail with another, but the average hen will yield to some of the following methods. We give the methods in the order of their preference:

Have a coop with slatted bottom suspended so it will swing easily. It can be hung from the roof in one of the houses, having it high enough so it can be reached easily and also so the confined hen can see the other hens at liberty. Put the setting hens in here that you wish to break, giving them plenty of feed and water. Give this coop a push every time you pass near it. Usually two or three days' confinement here will break them.

Another method is to confine the setting hens in a small dark pen, feeding them nothing. In this method put them in at night. Another method, put the setting hens in a small coop with no nest in it, feeding them plenty of corn and water. About five days is required by these methods. Another method is to place them in pens with several male birds only. Vigorous, active males are required and only put one hen at a time in the yard. This method depends very much on the number and vigor of the males in the yard.

The old method of "ducking" in water is not a very effective one. However, it will work on many hens if after the "ducking" they are put up in a coop and kept there for a few days, but the confining would doubtless break them, anyway.

Some poultrymen tie a red rag by a string a foot or two long to the tail of the setting hen and then turn her out to run her fright and setting off.

Egg Eating Hens.—It is a saying among poultrymen that an egg eating hen is worse than useless, as she will not only eat her own eggs, but also watch an opportunity to devour the products of other hens in the flock. They are usually started at this bad habit by breaking an egg, either soft shelled or because the nest has not sufficient nesting material in the bottom of it. When they once taste a broken egg they will soon learn to break the egg of their own accord. One trouble, too, is that egg eaters are usually to be found in small runs, without grass, and too often they are not given enough nitrogenous feed. One remedy is to feed more nourishing feed, more meat scraps, green bone, etc. Plenty of cabbage, lettuce, green alfalfa, etc., should also be fed. Cut down the corn and Kaffir corn and increase the meat and vegetables. Dark nests are also a good preventive of egg eating. The nests, to break egg eating, should be as dark as possible, with just a few rays of light entering at a slit at the back, and, of course, some light coming in where the hen must enter. The nest should be entirely covered and the opening where the hen enters faced toward the wall. Not being able to see the eggs while in this kind of a nest, the egg eater will leave without breaking the eggs. A remedy for egg eating is a mixture of cayenne pepper and hard boiled eggs, putting this in an egg shell so the hen will break the shell and eat the mixture. Another remedy is to keep a number of porcelain eggs lying loose in the nests, houses and yards. Gather the real eggs promptly, so that the hens will get tired of picking at the porcelain eggs and soon decide there is no use to bother. The point of the beak is sometimes trimmed to make it sore and prevent pecking at the real eggs.

It is important to keep the hens busy. Bones with some meat on them furnish meat food and keep the hens busy.

Feather Eating Fowls.—This trouble is more often seen among fowls that are in too close confinement with plenty of time on their hands and fed too much of some one food and not enough variety Fowls on the range, where they can select plenty of variety and keep busy taking exercise, are seldom bothered with this trouble. The first thing is to see if your fowls are fed a ration well balanced that does not want in some important food elements, as nitrogen, green feed, or even fat-forming feed, as corn, etc. Feeding on any one thing exclusively long will lead to this trouble. Get your birds to work early in the morning scratching for grain feed. Keep plenty of scratching feed to satisfy them and also to tire them well digging. Be sure to feed them plenty of cabbage, alfalfa meal—or, what is better, afalfa—lettuce, beets, turnips or other vegetables. Flour of sulphur is said to be a good remedy and three teaspoonfuls to every 25 fowls fed three times a week is about right in proportion. This should be fed them in the mash, and discontinued if it causes scouring. In the place where the feathers have been picked off, and the feathers near—usually on the neck or head—apply lard, in which has been dropped a few drops of carbolic acid and some bitter aloes. Five cents' worth of bitter aloes will make up several tablespoonfuls of lard. If the places where the feathers have been picked off are raw or bleeding, put on some carbolated vaseline. The beaks of any hens known to be feather eaters should be trimmed on one side, so as to make the beak slightly tender and thus discourage pulling and picking at feathers.

Preventing Cockerels Fighting.—Young cockerels raised together are not inclined to be "scrappy," but if kept in different pens for a few weeks, or in different colony houses, when put together in a bunch again, will frequently do considerable fighting. Also when taken off to shows they are more disposed to fight when brought back home. This can be largely prevented by placing the birds together in a darkened room for a day before turning them out. The room or pen should not be entirely dark, but just dark enough so they can get an indistinct view of each other. Another good way is to place an older and stronger bird with them, who will knock the cambatants apart and soon break them of the desire to scrap.

Preventing Flying.—If one wing only is clipped, trimming the main flight feathers about 2½ inches, the birds are unable to fly over a fence. If both wings are trimmed, the fowls can still fly some. Trimmed wings will not grow out as they will where the feathers are pulled out. If you trim off a wing the bird will not do to enter in the show room. One way to prevent flying over fences is to fasten a bent wire on one wing in such a manner that the wing cannot be straightened out. Fowls are not near so apt to fly over a fence where they cannot gauge the height. This is why a wire fence will turn fowls better than a board fence of the same height. A small wire stretched five or six inches above a board fence will prevent fowls flying over it. Another method is to tie the wing with a string or to hobble the foot so that the bird cannot jump up or make a spring with his feet. This is usually done by tying the string to the foot and up on the thigh or wing.

DETECTING OR CURING DISEASES.

Sick fowls show usually pale combs, wattles, etc., and are sluggish and not inclined to eat. Each disease also has special symptoms. As soon as any sign of sickness is noted the affected bird should be removed from the flock to a dry, warm place. Usually treatment does not pay and unless the birds are valuable it pays to kill for roup tuberculosis, cholera, etc., after the birds develop a clear case. Prevention emphatically beats treatment with fowls.

How to Prevent Sickness.—Sickness in fowls is much easier prevented than cured. Houses should be dry and free from draughts. plenty of exercise provided, food scrupulously clean, water pure, all drinking vessels and feeding vessels should be disinfected weekly. especially in warm weather, filth and droppings should be removed and houses kept clean and neat. Plenty of fresh air should be provided even in the coldest weather. Clean houses, clean feed, pure water, fresh air, plenty of exercise and well balanced feed will keep down and minimize the danger of disease.

Bumblefoot.—This is an abscess on the bottom of the foot caused by a bruise received probably by the fowl in jumping from a high roost. The treatment is to lance the abscess, remove the pus and then wash the wound thoroughly with a solution made of carbolic acid 1 part and warm water 5 parts.

Chicken-Pox or Sore-Head.—It is caused by an organism which is very contagious, the infection of which is usually from outside sources. The disease appears as small, yellow ulcers about the beak, nostrils, eyes and other parts of the head, hence the common name of "sorehead." Wash all affected parts with carbolic soapsuds and then apply vaseline to soften the sores.

Chicken Cholera.—This is a contagious disease of the bowels and may be observed by the characteristic green, frothy nature of the droppings. Treatment is generally useless. Kill all infected birds, burn or deeply bury, and disinfect the quarters thoroughly. Potassium permanganate or copperas in the drinking water of the well birds is a wise precaution.

Colds.—While these may not in themselves have much effect on the fowl, they often lead to something more serious and ought never to be neglected. Take equal parts of cayenne pepper, ginger and mustard, mix with lard or butter until a stiff paste is obtained. Roll these mixtures into little pellets and give to the affected bird by opening the mouth and dropping it down the throat. A single treatment will generally effect a cure. If not, repeat the dose later on.

Gapes.—In chicks this disease frequently destroys large numbers and is caused by tremadote worms in the windpipe. The trouble can be very easily identified, as the worms in the windpipe cause the bird considerable difficulty in breathing, causing it to open its mouth and gape, hence the name. Keep the fowls on fresh ground and do not allow their quarters to become filthy. A good remedy for the disease is to shut the affected fowls in a limited space and cause them to breathe air into which fine slacked lime is occasionally dusted.

Leg Weakness.—There are several things that causes a bird or chick to get weak in its legs. It affects young chicks and is often due to too strong feed. It is also caused by damp and crowded quarters and too many in a brooder. It is more noticeable in damp weather and cold, raw spring than in warmer weather. The first thing to do is to remove the cause. Give the affected fowls healthier surroundings and reduce the amount of both meat meal or scrap if you have been feeding them pretty heavily. If the affected ones have not already had plenty of exercise, make them scratch and dig for what they eat. Here is an English remedy that is said to be a cure for it. The dose is for grown fowls: Give pills composed of phosphate of lime 5 grains, sulphate of iron 1 grain, sulphate of quinine ½ grain, strychnine 1-16 grain. This quantity will make a dozen pills. Dose, 1 pill each day. For chicks give the amount of one pill to five chicks a month old. Feed in a soft mash.

Limber Neck.—As the name indicates, this disease is characterized by the limp condition of the neck. The fowls lose all control of the neck muscles and the head rests on the ground. It is caused by the fowls eating decayed flesh in which ptomaine has developed Treatment is rarely successful. The most effective and best treatment is to prevent it by being careful not to leave any decayed flesh where the fowls will have access to it.

A simple remedy for a limber neck is a few drops of turpentine given in some soft feed.

Lice.—Very often fowls taken to be sick are simply infested with lice. Dust baths should be provided, rosts, etc., sprayed, and the birds themselves given a good dusting with insect powders, either those put up by some reliable company or they may be obtained in bulk at the stores. In dusting for lice, work the powder carefully into the feathers, especially along the back and on the sides. For small chicks, do not use powders, but grease the back of the head, as advised under "Some Little Chicks Suggestions." Houses, roosts, nests, coops, etc., should be disinfected as directed elsewhere. (See "Recipes and Remedies" for some good disinfectants and louse killers.)

Mites.—This pest lives on the roosts, walls, nests, etc., in daylight and sucks the blood of the fowl at night. Use the louse killers and disinfectants on walls, etc., as directed elsewhere.

Roup.—This term is used to apply to a number of diseases affecting the throat. What is now generally recognized as roup is diphtheria or diphtheritic roup. And that which is ordinarily called roup, which does not seriously affect the throat, is influenza or heavy cold. Make a swab of cotton and tie it on the end of a stick and swab the mouth with hydrogen peroxide. For the mild form, wash sores and discharges from the nostrils with a 2 per cent solution of carbolic acid and give some stimulant, such as ginger or red pepper, in the food.

Buildings, roosts and runs should be thoroughly disinfected with a reliable disinfectant. In severe cases, one of the following should be injected in the mouth and nostrils, carefully washing any sores found. The fourth item should be used in the drinking water: (1.) Two per cent solution of some coal tar disinfectant. (2.) Two per cent solution of carbolic acid. (3.) Peroxide of hydrogen and water, equal parts. (4.) One grain permanganate of potash to an ounce of water. (5.) Kerosene oil mixed with equal parts of lard or olive oil.

Here is a remedy that is said to be a good one for roup. Give a pinch of the mixture for each fowl. It should be mixed with the feed. Hyposulphite of soda, 50 grams; salicylate of soda, 50 grams; pulverized ginger, 200 grams; pulverized yellow gentian, 200 grams; pulverized sulphate of iron, 100 grams.

Scaly Legs.—This disease is caused by a small parasite on the feet and legs of the fowls. If not treated, the legs will soon become very much roughened, enlarged and knotty and the natural color ruined. The legs of the affected fowls should be first carefully washed in warm soapsuds so as to clean off all the rough outside formation, but care should be taken not to break through and cause the feet and legs to bleed. After washing and drying thoroughly, rub carefully with a mixture composed of the following: Five parts kerosene, 1 part crude carbolic acid, 1 part sulphur. It will take two applications three days apart to effect a cure. Roosts, houses, nests etc., should be well sprayed or painted with good disinfectant.

Tuberculosis.—This is sometimes called "going light," as the birds lose weight. The birds with this disease stand around or "mope" about. Feed soft food and an abundance of fresh meat, also give all the chopped onion and green food they will eat.

HOW TO PRESERVE EGGS.

"Preserved eggs may be substituted for fresh ones in many cases with profit. They may be scrambled and used in omelets; also for baking various cakes which do not require beaten whites. As a rule, they are the equivalent of fresh eggs in any food where the yolk is broken; but only when specially preserved and when kept not too long are they suitable to serve fried.

"The preserving material seals up the pores in the shell and thus prevents the entrance of bacteria and air, as well as evaporation and consequent shrinkage of the egg contents. The old method of greasing the shell to make eggs keep better depended on this fact. Such egg cannot be boiled because the impervious shells do not permit the escape of the enclosed air, which expands when heated and bursts open the egg. By serving the commoner purposes the preserved egg economizes the fresh egg, for which there is an ever-increasing demand."

Dry Packing Method.—To keep eggs a short time only, the usual methods of packing are sufficient. For this purpose they are embedded in some fine material, such as dry bran, oats, sawdust or salt. Care must be taken that the packing material is perfectly dry and free from must. There is always danger of losing the eggs by the growth of mold on the inside of the shells. A better way is said to be the use of egg shells. These are arranged in a cool, dry place and are provided with holes so that the eggs may be stood on end. Handled in this way, eggs are said to keep better when packed. Preserving in some solution is, however, a much safer method for general use.

General Precautions.—Attention should be paid to the following.
1. That eggs perfectly fresh only be used, not over two days old.
2. That the eggs should throughout the whole period of preservation be completely immersed.
3. Better results can be had by using from hens separated from the male birds.

Although not necessary to the preservation of the eggs in a sound condition, a temperature of 40 degrees F. will materially assist toward retaining good flavor or rather in arresting that "stale" flavor so often characteristic of packed eggs.

Limewater Method.—The solubility of lime at ordinary temperature is one part in 70 parts of water. Such a solution is termed saturated limewater. In pounds and gallons, this means 1 pound of lime is sufficient to saturate 7 gallons of water. However, owing to impurities in commercial lime, it is well to use more than is called for in this statement. It may not, however, be necessary if good, freshly burned quicklime can be obtained, to employ as much as was at first recommended, namely, 2 to 3 pounds to 5 gallons of water. With such lime as is here referred to, one could rest assured that 1 pound to 5 gallons (50 pounds) would be ample, and that the resulting limewater would be thoroughly saturated. The method of preparation is simply to slake the lime with a small quantity of water and then stir the milk of lime so formed into 5 gallons of water. After this mixture has been well stirred for a few hours it is allowed to settle. The supernatant liquid, which is now "saturated" limewater, is drawn off and poured over the eggs, previousy placed in a crock or watertight barrel.

As exposure to air tends to precipitate the lime (as carbonate) and thus weaken the solution, the vessel containing the eggs should be kept covered. The air may be excluded by a covering of sweet oil, or by sacking upon which a paste of lime has been spread. If after a time there is any noticeable precipitation of the lime, that limewater should be drawn or siphoned off and replaced with a further quantity of newly prepared.

Limewater preserved eggs will keep well and are serviceable for all purposes excepting to fry, the yolks not holding up well and the eggs being apt to become mussy. There is a great tendency for the whites to become watery, but this does not render the eggs unwholesome. They are just as serviceable for baking and for other purposes as fresh eggs, excepting that the whites cannot be beaten. Limewater is regarded as the best preservative.

Water Glass or Sodium Silicate Method.—This is diluted with from 10 to 20 parts of water, but even greater dilutions will serve when the eggs are to be kept a short time only. The stronger the water glass solution, the less apt the yolks are to break when fried. On the whole, solutions 2 to 5 per cent (two pounds to five gallons of water) have given better results than stronger solutions.

Hot Water Method.—Hold fresh eggs in boiling water while you count six, using a wire basket. Let them dry, cool and pack in a layer of oats, small end down. Put a layer of oats, then one of eggs. It is claimed that eggs packed this way will keep several months and be fresh as when first put down. They should be kept in a cellar or some cool place.

PACKING EGGS FOR SHIPMENT.

I find that the egg buyers would rather pay a little more express and be sure of getting their eggs sound than to have them packed too light and receive broken eggs. I pad the bottom and sides of a half bushel market basket well with excelsior, wrap each egg carefully in a piece of paper 6x9 inches, then wrap the egg and paper in a bundle of excelsior. I press them down firmly so that they will not jar, and I then pack plenty of excelsior and sew down a cloth on top of the basket good and tight. I have had very few complaints about hatches and scarcely any about broken eggs. I put a postal card in each basket, asking my customers to report broken eggs, bad hatches, etc.—H. A. Sibley.

Recipes and Remedies

POPULAR ENGLISH REMEDY FOR ROUP.

The following is said to be the recipe for making Vale's roup pills, a popular English remedy: Hydrastin, 2 grains; sulphate of iron (dried) and sulphate of copper, 3 grains of each; powdered capsicum, 12 grains; oil of copaiba, 20 drops; Venetian turpentine and calcined magnesia, of each enough to make 24 pills. Dose for adult fowl, one or two pills night and morning.

MAKING A GOOD LICE KILLER.

A good liquid lice killer is made by dissolving in ordinary kerosene all the crude napthaline flakes it will take up. The solution is an excellent disinfectant for use about the poultry house, as well as a lice killer. Painted on the dropping boards and roots, it will destroy and prevent red mites, and will also kill disease germs.

AN EGG TONIC.

Four ounces of soda (common baking soda), 1 ounce of powdered charcoal and 1 ounce each of the following: Carbonate of iron, mandrake, ginger, pulverized gentian root, flour of sulphur and black antimony. In feeding add 1 teaspoonful of this to every quart of mash feed given. When the fowls are doing well, once a week is often enough to give this tonic.

FLOCK TONICS.

This tonic may be given with excellent results to fowls moping and not doing well. It is also good to feed during the molt: Linseed meal, 4½ pounds; powdered nux vomica, 1 ounce; powdered ginger, 2 ounces; cayenne pepper, 2 ounces. Give a teaspoonful to each 10 hens morning and night in mash for two days, then miss a day.

Here is another: Cayenne, liquorice and table salt, 2 ounces of each; gentian and bone meal, 4 ounces of each; fenugreek and brown sugar, 8 ounces of each. These must be reduced to powder form in a mortar with a pestle, (as a rule) ere they can be mixed together, which they should be in a thorough manner. A tablespoonful of this spice to every 10 or 12 hens twice or thrice a week is a good allowance. It should be mixed with the breakfast soft feed, and is best used in late autumn and throughout winter.

DOUGLAS MIXTURE.

The Douglas Mixture was extensively used by poultrymen years ago as a tonic and astringent, but its value was severely criticized by several writers and the mixture lost repute.

The Douglas Mixture is made and used as follows: Sulphate of iron, ¼ pound; sulphuric acid (pure), ½ ounce; water, 1 gallon.

Place the sulphate of iron and the acid in a strong earthenware pot; pour on the water and stir well with a stick. Cover the mixture with a wooden cover and leave for a day. Then run it off into bottles and cork well and seal. Dose—A teaspoonful to each half pint of drinking water, or in similar proportions in the water used for mixing the food, every third or fourth day. The drinking vessel and others holding the mixture should not be metal.

The following formula is more satisfactory: Sulphate of iron, ¼ pound; aromatic sulphuric acid, 4 ounces, and 1 gallon of water. Mix and use as advised. The aromatic sulphuric acid is a solution of the acid in alcohol, with tincture of ginger and spirits of cinnamon added, and is preferable for internal use.

A GOOD ROOST LOUSE PAINT.

By painting roosts, walls, etc., of poultry houses with the liquid louse killer given below, the fumes will tend to keep down the rapid increase of insect pests.

The fumes are strong and the fowls may not like to go inside the building. Do not close doors or windows, as the fumes will hurt the flock if you do. The paint should be put on in the morning, closing all doors and windows for half a day, keeping the fowls out of the house. Open all doors and windows in the afternoon so the house may be aired out by night.

One gallon crude kerosene oil; ½ pound crude carbolic acid; ¼ pound carbon disulphide; 1 gill pine tar.

Stir the tar in last and in small quantities. Common kerosene oil may be used, also refined carbolic acid, but these refined articles are more costly and not so effective. The pine tar is largely to make the mixture stay on well. The carbon disulphide is very powerful and care must be exercised not to inhale its fumes while mixing. The mixture should be kept in a stone or glass vessel and stopped up airtight or it will soon evaporate and lose its strength.

WHITEWASH GOOD AS PAINT.

All poultry buildings, nests, roosts, feed hoppers, etc., should be painted at the beginning of each year. The paint permanently fills all cracks, crevices and rough places where lice and mites would hide. It renders the surface smooth, so that disinfectants and lice killers sprayed or painted on the nests, feed hoppers, etc., are not absorbed by unpainted wood. Thus paint is cheaper than unpainted wood, to say nothing of the looks and longer services of painted material. It is best to paint with a cheap barn or house paint mixed with oil, but if you do not feel that you can afford it, what is known as "government whitewash" will last nearly as well as paint, and looks good, also.

This whitewash, used by the government for lighthouses, etc., is made as follows: Slake in boiling water one-half bushel of lime. Strain so as to remove all sediment. Add 2 pounds of sulphate of zinc, 1 pound of common salt and ½ pound whiting, thoroughly dissolved. Mix to proper consistency with skimmilk, if possible, and do not use water. Stir in thoroughly a half pint of liquid glue and apply the wash while hot. It may be colored if desired by using yellow ochre, ultramarine blue, lamp black, etc.

POWERFUL DISINFECTANT AND LOUSE KILLER.

There is no use paying from $1.50 to $3.00 a gallon for disinfectant and louse killer when you can prepare your own for a few cents a gallon. The one given below will not only kill all germs, will also keep down mites and if sprayed on the poultry in the morning will keep down scaley leg and lice. It should not be sprayed on the poultry oftener than every other week. Commercial cresol, the basis of this louse killer, sells for 25 to 30 cents a pound. The cresol should be handled with great care and if it gets on the hands or skin should be washed off immediately with clean water, as it burns. The mixture is prepared in the following manner:

Measure out 4 quarts of raw linseed oil in a 4 or 5 gallon stone crock; then weigh out in a dish 1¾ pounds of commercial potassium hydrocide or caustic potash, which may be obtained from any druggist at a cost of from 10 to 15 cents a pound. Dissolve this caustic potash in 1 pint of water; let it stand for at least three hours until the potash is completely dissolved and the solution is cold; then add the cold potash solution very slowly to the linseed oil, stirring constantly. Not less than five minutes should be taken for the adding of this solution of potash to the oil. For five hours after mixing the oil and potash mixture (soap) should be stirred thoroughly about once every hour and then left standing for ten or twelve hours. By the expiration of that time saponification should be complete. The soap should then be stirred and broken up into small pieces and 5¼ quarts of commercial cresol should be added. The soap will slowly dissolve in this cresol. It may take two days for complete solution to be effected. The length of time taken in dissolving will depend on the condition of the soap, which in turn varies with different lots of linseed oil. When the soap is all dissolved, the solution, which is liquor cresolis compositus or cresol soap, is then ready to use. The cresol soap will mix in any proportion with water and yield a clear solution. Three or four tablespoonfuls of the cresol soap to each gallon of water will make a satisfactory solution. This solution may be applied through any kind of spray pump or with a brush. Being a clear, watery fluid, it can be used in any spray pump without difficulty. For disinfecting brooders or incubators which there is reason to believe have been particularly liable to infection with germs of white diarrhea or other diseases, the cresol may be used in double the strength given above and applied with a scrub brush in addition to the spray.

www.ingramcontent.com/pod-product-compliance
Lightning Source LLC
Chambersburg PA
CBHW062336220526
45469CB00008B/2735